INTERNATIONAL CENTRE FOR MECHANICAL SCIENCES

COURSES AND LECTURES No. 201

Ro. man. sy. '73

FIRST CISM - IFToMM SYMPOSIUM
5 - 8 September 1973

ON THEORY AND PRACTICE

OF

ROBOTS AND MANIPULATORS

VOLUME II

UDINE 1974

Springer-Verlag Wien GmbH

ISBN 978-3-662-39349-9   ISBN 978-3-662-40393-8 (eBook)
DOI 10.1007/978-3-662-40393-8

5. ARTIFICIAL INTELLIGENCE

THE CONCEPT OF THE STRUCTURE OF HIGHEST LEVELS CONTROL IN ROBOT MANIPULATORS

M. V. ARISTOVA, Cand. of Engng Sci

M.B. IGNATIEV, Dr. of Engng Sci
Leningrad Institute of Aviation
Instrument Making Ul. Gerzena 67,
Leningrad, USSR.

(*)

РЕЗЮМЕ

Для построения высших уровней управления роботами-манипуляторами ие-обходимо изучить структуру его поведения.Следует вычленить основные струк-турные этого поведедения, число которых должно быть не больше нескольких десятков и которые позволяли бы описать и оргонизовать сложное поведение обозримым образом.Это облегчить разработку ряда промышленных и исследова-тельских роботов и, как надеются авторы, внесет определенный порядок в проблему искусственного интеллекта.

Виделенные структурные единицы предлагается затем положить в основу некоторой формальной теории, например, теории первого порядка.Рассматривая сложные задания как потенциальные теоремы этой теории, можно с помощью ав-томатического доказателя теорем на ЭВМ получать цепочки элементарных дей-ствий, приводящих к цели.Для удобства общения с роботом целесообразно ис-пользовать ограниченный естественный язык ЯДРО- LAROM , близкий по термино-логии к сфере деятельности робота.Связь его внутренним языком принятия ре-шения осуществляется через транслятор.

(*) All figures quoted in the text are at the end of the lecture.

1. Introduction

Robot-manipulators are assigned to execute different tasks in the complex environment. Nowadays it is worth of speaking about three levels of control [1]. The lowest level has mechanical motion at the output, and at the input it has command signals for automatic drive. The second level produces signals for the first one on the basis of general orders such as: go to the point, go along the straight line, take an object, find an object, etc. Now the second level structure acquires definite features, and it is the bases of supervisory control [1]. Highest levels have not yet precise structure, it is in an embrionic state. Design of these highest levels of control is related with great difficulties which explains the importance of departing from the customary method of coordinates.

On the highest levels we deal with robot behaviour in the world of things and if we tried to set their coordinates we would get vast description which could not be processed, for the design of highest levels of control, it is necessary to study the robot behaviour structure which reminds one of literary script. It is necessary to single out the main structural units of its behaviour, the number of which should not be more than several dozen and which could allow one to describe and organize complex behaviour by some known method. So it is natural to use methods of literary structuralism.

A Soviet folklorist V.Y. Propp as early as 1928 [2] tried to find constant unchangeable elements meeting in any fairy-tale. An approach like this made it possible to describe all fairy-tales as a single fairy-tale — the same problem posed for robot control is to pick out main elements of behaviour with systemtic analysis of numerous situations of complex behaviour in different environments thereby to set the structure of the robot in its highest levels. V.Y. Propp showed that the undecomposable unit of the fairy-tale is not content or a motive, but a function, moreover, the function implies an action of the character determined from the standpoint of its significance for the development of events. Functions of characters are invariants of the fairy-tale. Their number is restricted — V.Y. Propp distinguished 31 functions.

Taking into consideration methods of semiotics it is possible to pose the same problem for algorithms of robots control to reveal their main compatible functions. It will simplify the designing of a number of industrial and research robots, and, as the authors hope, will create a definite order in the problem of artificial intelligence. At the present time this problem needs its own Mendeleev's table and only with its aid will it be possible to put in order the obtained results and determine the problem at hand in the research on artificial intelligence and integrated robots.

2. The first order theory for robot automatic control.

One of the possible means of robot automatic control on the third level is to take advantage of the formal theory of mathematical logic. The presence of quantifiers and predicates makes it possible to speak about the theory as the first order theory, which axioms are divided into two groups: logical and proper axioms [4]. Simulation of environment is done by means of well-formed formulas (statements) in respect to the location of available objects and the robot in a certain present situation which may be considered as the proper axioms of the I group. Their peculiarities lie in the fact that they do not remain invariable during the robot performance and they should be corrected as the results of every task execution. The proper axioms of the II group should establish a certain correspondence between robot transference and objects location in the environment and according to it they contain a set of elementary functions of robot performance [5]. In this case a definite task expressed in the form of the corresponding statement of the theory is potentially capable of execution only in the case when it is the consequence of the above mentioned axioms, i.e. the theorem of this theory. In the considered application it is necessary to get a sequence of elementary functions of robot action, resulting in the fixed aim.

So two problems are set: selection of the corresponding system of axioms and adaptation of a computer program for proving the automatic reception of the elementary functions sequence of robot performance.

In the formulation of the proposed theory the following symbols take place:

1. Punctuation marks: round brackets, point, comma.

2. Propositional links: negation - -, conjunction - &, disjunction - V and implication ⊃ .

3. Variables: small letters - o, s, u, v, w, x, y, z with indexes or without them. Variables of the theory are interpreted by several ranges: by the set of singular volumes of manipulation space {u, v, w}; by the set of objects {o} and by the set of situations {s}, where the situation means the definition of environment state, in which it is the distinguished aspects interesting us during the performance of the task, for instance, location and shape of the object but not its colour of material [6].

4. Functions: small letters - a, b, c, f, g, h, m, p with indexes or without

them. Symbols for 0-argument functions are individual constants, in particular a_1, $a_2...a_n$ - are names of objects, b_0, b_1,...b_k - are names of manipulation space points; c_0 - serves for the indication of the present situation, transformation of which will later on result in achievment of the goal; m - denotes the robot manipulator.

At the situation set was defined the following functions of one argument: p(s) - a function interpreted by elementary motion of the manipulator to open the grasp and h(s) - the function interpreted by the motion to shut the grasp. Their meaning is the situation proceeding immediately after execution of given real actions. At the same set $\{s\}$ the following functions of the arguments are established: f(u, v, s) the meaning of which gives a situation coming after the fulfilment of the action by the robot going from the point u into the point v, and the function g (w, w_1, s) the meaning of which is the situation determined by the conditional expression:

(1) $g(w, w_1, s) = \underline{if}\ R(w,\ w_1, s)\ \underline{then}\ s\ else\ g(t(w), w_1,\ f(w, t(w), s))$

Here we get a new function t(w). It is the function of coordinate transformation of robot location depending on a specific type of trajectory of its motion symbol. R is the referred to predicate and it is described below.

5. Predicates: capital letters — B, C, H, K, M, P, R, Q with indexes and without ones. In the given interpretation predicate letters are arranged in accordance with some relations at the given sets. For instance, $K_1(o)$ means that the object belongs to the class of the cubes. 3 place predicate M (a_1, b_{10}, c_0) means that an object a_1 is in the point b_{10} in the present situation c_0. Mentioned above 3-place predicate R (w, w_1, s) is interpreted in the following way, it takes the meaning "truth" if the point w_1 belongs to certain (beforehand fixed ϵ - locality of point w and meaning "false" - in the opposite case. 2-place predicate P (v, w) means that points v and w coincide. It is clear that the meaning of predicates is logical constants "truth" and "false". As to functions of n arguments and n-place predicates it is performed by a rule on the base of which their arguments create strictly ordered sequences, for example, ordered three.

6. Quantifiers: expression V, called universal quantifier (i.e. for all the variables) and E, called existential quantifier (i.e. there is such a variable, that...).

As it has been mentioned above, the theory includes an arbitrary but finite set of axioms, characterising the position of the robot-manipulator and objects in the present situation c_0. For instance, if manipulation space contains two cubes and a prism in the known points, the following axioms of the I group are introduced:

$$K_1(a_1) \ \& \ M \ (a_1, b_{10}, c_o); \quad \neg \ P \ (b_{10}, b_{20});$$

$$K_1(a_2) \ \& \ M \ (a_2, b_{20}, c_o); \quad \neg \ P \ (b_{20}, b_{30});$$

$$P_1(a_3) \ \& \ M \ (a_3, b_{30}, c_o); \quad \neg \ P \ (b_{10}, b_{30});$$

From the proper axioms of the II group introduced into the theory, we mention below the three main axioms.

Axiom of transference:

$$\forall u \forall v \forall w \forall o \forall s \ (M(o,v,s) \ \& \ M \ (m,u,s) \supset M \ (o,w,p(f(v,w,h(p(f(u,v,s))))))^{(*)} \ M \ (m,w,s_1)) \tag{2}$$

The meaning of this axiom lies in the following: if arbitrary object o is in the point v and the manipulator is in the point u at some situation s then for the transfer of the object o, into the arbitrary point w it is necessary to transfer the manipulator from the point v, to open the grasp for the taking of the objects, shut the grasp and then remove the manipulator with the grasped object into the point w, where the grasp is opened.

Axiom of independence:

$$\forall u \forall v \forall v_1 \forall w \forall o \forall s \ (M(o,v_1,s) \ \& \ M(m,u,s) \ \& \ \neg \ P(v,v_1) \supset M \ (o,v_1,$$
$$p(f(v,w,h(p(f(u,v,s)))))) \tag{3}$$

This axiom means that the position of any object is not influenced by transference of other object or the very manipulator, if these motions do not touch a given object.

Axiom of motion. For the decision of the tasks, related with motion of the robot manipulator along a trajectory on the plane, the following axiom is introduced.

$$\forall v \forall w \forall w_1 \forall s \ (M(m,w,s) \supset M(m,v,g(w,w_1,s)) \ \& \ R \ (v,w_1,g(w,w_1,s)) \tag{4}$$

(*) designate the third argument by s_1.

A function g, as it has been noted, is determined by means of conditional expression [1] and it is dependent on the function f (v, w, s), interpreted by the elementary motion going from the point v into w. For the computation of the consequent meanings of this function we make the following scheme of primitive recursion:

$$(5) \quad \begin{cases} g\ (0,w,w_1,s) = s \\ g\ (n',w,w_1,s) = g(n,t(w),w_1,\ f\ (w,t(w),s)) \end{cases}$$

So function g creates a chain of situations, distinguished from one another only by the location of the robot-manipulator which moves along the trajectory defined by the function t(w).

It is known that the function $g(w, w_1, s)$ will fall under the category of recursive only in the case when the functions t(w) and f(u, v, s) which it is dependent on are recursive. In reality it is not difficult to make recursive description for functions f, p and h. As to a recursive description of the function t it should be noted that the authors made a recursive description for the function of robot transference along the arbitrary straight line, hyperbola, parabola, and arc of a circle.

All the three axioms are closed formulas and they are true for the selected interpretation. Tasks which may be carried out by the robot-manipulator are formed as well formed statement W, which should be the consequence of the described axioms A_1, ..., A_n. A procedure of automatic proof of the theorem W comes to the proof of inconsistency of the list A_1, ... $A_n \neg W$. For the reduction of time expanded on the proof related with the number of variants we used the principle of resolution offered by american logician J.A. Robinson [7]. However any automatic prover may answer (in the affirmative way or in the negative way) only the question whether the offered task may be performed.

The second problem covers the definition of the consequence of elementary functions of action inherent in axioms and leading to the task fulfilment. Since it usually comes to the discovery of variables meanings, situated under the existential quantifier it is proposed to add to the statement negation an auxiliary predicate ANSWER $(x_1,$ ..., $x_n)$ from the variables mentioned above as C. Green [8] does in question-answering systems. An expression ANSWER (c_1) V...V ANSWER (c_k) obtained in this case a result of the proofis a false statement and thereby it affirms that initial task W is the consequence of the list $A_1,...A_n$. Worked out variable meaning with the predicate ANSWER is one of the examples of the situation sought s (as s was under the existential quantifier) made of the combination of elementary actions leading to the aim.

3. An example of automatic construction of the plan of robot actions

Let's consider a task consisting of the transference of the object a_1 into the point b_1, and of the object a_2 into the point b_2 which takes in the first order theory a form of statement (theorem):

$$Es \ (M(a_1,b_1,s) \ \& \ M(a_2,b_2,s) \ \& \ M(m,b_2,s))$$

For the proof let's take the proper axioms of transference and independence which after elimination of quantifiers and transformation into the conjunctive normal form (c.n.f.) become the following:

1. $\overline{M}(o,v,s) V \ \overline{M}(m,u,s) V \ M(o,w,p \ (f(v,w,h(p(f(u,v,s))))))$
2. $\overline{M}(o,v,s) V \ \overline{M}(m,u,s) V \ M(m,w,p \ (f(v,w,h(p(f(u,v,s))))))$
3. $\overline{M}(o,v_1,s) V \ \overline{M}(m,u,s) V \ P(v,v_1) V \ M(o,v_1,p(f(v,w,h(p(f(u,v,s)))))))$

Let s be in the present situation c_0, the object a_1 is in the point b_{10} the object a_2 - in the point b_{20}, and the robot - manipulator - in the point b_0. This information is used in the form of the following axioms.

4. $M \ (m, \ b_o, \ c_o)$
5. $M \ (a_1, b_{10}, \ c_o)$
6. $M \ (a_2, b_{20}, \ c_o)$
7. $P \ (b_{10}, b_{20})$
8. $P \ (b_{20}, b_1)$

Transformation into c.n.f. of the assumed theorem negation causes a disjunct of the following form.

9. $\overline{M} \ (a_1, b_1, s) V \ \overline{M}(a_2, b_2, s) V \ \overline{M}(m, b_2, s) V \ ANSWER \ (s)$

Below is given a proof with the setting of a satisfying meaning s.

10. $\overline{M}(o,v,c_o) V \ M(o,w,p(f(v,w,h(p(f(b_o,v,c_o)))))) \ (*)$

p.1.4	$\{b_o/u, \ c_o/s\}$
p.2.4	$\{b_o/u, \ c_o/s\}$

11. $\overline{M} \ (o,v,c_o) \ v \ M \ (m,w,s_1)$
12. $\overline{M} \ (a_1,w,p(f(b_{10},w,h(p(f(b_o,b_{10},c_o)))))) \ (**)$

p.5.10	$\{a_1/o, \ b_{10}/v\}$
p.5.11	$\{a_1/o, \ b_{10}/v\}$

13. $M \ (m,w,s_2)$

(*) designate the third argument by s1

(**) designate the third argument by s2

14. $\bar{M}(o,v,s_2) V\ M(o,w_1,p(f(v,w_1,h(p(f(w,v,s_2))))))$ (*)

p.1.13 $\{w/u,\ w_1/v,s_2/s\}$

15. $\bar{M}\ (o,v,s_2)\ V\ M\ (m,w_1,s_3)$

p.2.13 $\{w/u,\ w_1/v,s_2/s\}$

16. $\bar{M}(o,v_1,c_o) V\ \bar{M}(m,b_o,c_o) V\ P(b_{10},v_1) V\ M(o,w_1,p(f(v_1,w,h$
$(p(f(w,v_1,s_3)))\ldots)$ (**)

p.3.14 $\{b_{10}/v,b_o/u,v_1/v\}$

17. $\bar{M}\ (o,v_1,c_o)\ V\ P\ (b_{10},v_1)\ V\ M\ (o,\ w_1,s_4)$

p.4.16

18. $P(b_{10},b_{20}) V\ M(s_2,w_1,p(f(b_{20},w_1,h(p(f(w,b_{20},s_2))))))$ (***)

p.6.17 $\{s_2/o,b_{20}/v_1\}$

19. $M\ (s_2,w_1,s_5)$ p.7.18

20. $\bar{M}(o,v_1,c_o) V\ \bar{M}(m,b_o,c_o) V\ P(b_{10},v_1) V\ M(m,w_1,s_4)$

p.3.15 $\{b_{10}/v,b_o/u,v_1/v\}$

21. $\bar{M}\ (o,v_1,c_o)\ V\ P\ (b_{10},v_1)\ V\ M\ (m,w_1,s_4)$

p.4.20

22. $P\ (b_{10},b_{20})\ V\ M\ (m,w_1,s_5)$ p.6.21 $\{s_2/o,b_{20}/v_1\}$

23. $M\ (m,w_1,s_5)$ p.7.22

24. $P(v,w) V\ \bar{M}(m,u,s_2) V\ \bar{M}(a_1,w,p(f(v,w_1,h(p(f(u,v,s_2)))))$

p.3.12 $\{a_1/o,w/v_1\}$

25. $P\ (v,w)\ V\ \bar{M}\ (a_1,w,p(f(b_{20},w_1,h(p(f(w,b_{20},s_2))))))$

p.13.24 $\{w/u\}$

26. $\bar{M}(a_1,b,p(f(b_{20},b_2,h(p(f(w,b_{20},s_2))))))$(****)$V M(m,b_2,s_6)\ V$
V ANSWER (s_6) p.9.19 $\{b_2/w_1\}$

27. $\bar{M}\ (a_1,b_1,s_6)\ V$ ANSWER (s_6) p.26.23 $\{b_2/w_1\}$

28. $\bar{M}\ (a_1,b_1,p(f(b_{20},w_1,h(p(f(b_{20},p(f(b_{10},b_1,h(p(f(b_o,b_{10},c_o}\ldots)$

p.8.25 $\{b_{20}/v,b_1/w\}$

29. ANSWER $(p(f(b_{20},b_2,h(p(f(b_1,b_{20},p(f(b_{10},b_1,h(p(f(b_o,b_{10},c_o)\ldots)$

p.28.27 $\{b_1/w,b_{20}/v,b_2/w_1,b_1/u,b_{20}/v\}$

(*) designate the third argument by s_3
(**) designate the third argument by s_4
(***) designate the third argument by s_5
(****) designate the third argument by s_6

So the situation sought is determined by means of the argument of the predicate ANSWER. For the construction of required elementary action sequence it is necessary to display an expression for s placing meaning on functions p,h and f. As the result we get a chain: OPEN; SHUT; GO (b_{20}, b_2); OPEN; GO (b_1, b_{20}); OPEN; SHUT; GO (b_{20}, b_2); OPEN. In accordance with the name of objects and points of their new location it is an automatically created necessary sequence of actions without human intervention. After realization of this sequence on the low level of control old axioms of the I group incompatible with robot's and objects' locations are substituted by new ones:

$$M \ (m, \ b_2, \ c_o)$$
$$M \ (a_1, b_1, \ c_o)$$
$$M \ (a_2, b_2, \ c_o)$$

It is known that during the transformation necessary for the proof of a statement W by means of a computer (corresponding, for example, to the task related with transference of all the cubes into the point b_1) having a number of variables under the universal quantifier there appears uncertain Skolem functions of these variables. In accordance with the fact that these functions have not replaced the universal quantifier by some combination of existential quantifiers by means of the relation

$$VoR \ (o) \ = \ Eo_1 Eo_2 \ldots Eo_n R(o) \tag{6}$$

It is possible to prove that this relationship is correct on a finite model of environment, having n objects with the property R. Due to the above mentioned task on collecting all the cubes taken in the theory form:

$$EsVo(K_1 \ (o) \ \& \ M \ (o, b_1, s) \ \& \ M \ (m, b_1, s))$$

it succeeded to come to the task of the collection of n objects into one point and to obtain the sequence of sought for elementary robot actions.

In addition to the list of axioms, an automatic theoremprover should serve, firstly, as a programme which could form axioms of the I group during uncertain situations; secondly, as a programme which could manage to select a list of corresponding axioms to a given task; thirdly, as some subset of languages close to the natural language of higher level on the basis of which it will be possible to formulate a task referred to different professional spheres of activity. Moreover a created plan of actions (in the form of the sequence of function) leads to success only in the case when these axioms strictly correspond to the environment.

4. A language of robot motions   ЯДРО   (LAROM)

Some words about the possible construction of natural language of the higher level. One of the variants of such language is described in [1, 3]. As to form it represents limited natural language and it is constructed in the form of orders, which are analogues to orders issued by a human operator. Creation of the universal language with the orientation to its usage is complicated and time-consuming work. The available variant of language is destined for the description of tasks related with motion of specific objects into a given place, selection of objects, building of simple constructions of unify blocks, motion to the given point, motion along the given trajectory, and so on. This language is called LAROM - a language of robot motions.

Its peculiarity is the introduction of semantic words, which part is similar to delimiters of ALGOL. Words belong to main symbols of language and they are indivisible units. A set of all words in the language forms its vocabulary. Words of language are not further defined, but they should meet some requirements resulting from the specific features of information furnishing to a concrete computer, for example, case-endings have no meaning. It is possible to put down words in the algorithm without abbreviations. An intelligent unit (analogue of sentences in speach) is an order in this language by means of which manipulator control is done. In the present variant there are seven groups of orders the number of which may be increasing while the language is extended.

$$< order > : = < go > | < find > | < carry > | < construct > |$$
$$< simulate > | < execute > | < stop >$$

When any order has been carried out a system aswers a operator "ready". If for some reason the present order could not be carried out, operator attention is attracted by the signal "SOS". For the possibility of compatible execution of some orders they are provided with numbers (the whole two-digit numbers) situated in front. Then execution of the set of orders, they are done after the order "execute" in the succession of given numbers.

For the translation from the language LAROM in the language of the first order theory it is necessary to have a syntactical translator similar to that which is used for the translation from one natural language into another. At this stage it is possible to choose axioms necessary for the proof of a theorem-task falling under the definite category of orders, it significantly reduces the number of substitutions, when an automatic proof is done.

The advantage of the language LAROM is the visual method of task setting; unambiguity of any order record which simplify translation; relation with the language of the lower level; possibilities of language extension in the case of control of the robot, taking part in different processes.

5. Conclusion.

The considerations above cover only some aspects of the structure design of the robot control highest levels. The design of highest levels control is an important condition for successful applications of robots for execution of complicated tasks.

For the utilization of the proposed theory on the control of robot performing more complicated tasks, it is necessary to introduce additional proper axioms of the II group which describe the essence of the task by means of the formal method. So the extension of the robot activity sphere is just related to the addition of new corresponding axioms and does not demand reprogramming of the theorem-prover by means of the actions by which a course of the robot is carried out.

As advantages of the proposed approach we may refer to the following. In the first place, by means of the automatic theorem-prover there is actually done initial simulation of the task by means of the computer without its actual execution on the mechanical level. The success of this stage determines all the subsequent actions. In the second place, introduction of the environment model and information related with the robot in the form of the first order theory statements permit the use of any universal theorem-prover, designed for another purpose. In the third place, since the influence of robot design peculiarities appears only on the lower control level, the described part of the system may be used for the (robot) control of different designs.

REFERENCES

[1]   ИГНАТЬЕВ М.Б., КУЛАКОВ Ф.М., ПОКРОВСКИЙ А.М.- Алгоритмы управления
      роботами-манипуляторами, Изд.Машиностпоение, Ленинград, 1972.

[2]   ПРОПП В.Я.- Морфология сказки, Изд. Academia,Ленинград, 1928.

[3]   АРИСТОВА М.В., ИГНАТЬЕВ М.Б.- О разработке алгоритмического про-
      блемно-орентированного языка для управления манипуляторами.
      "Труды Ленинградского Института Авиационного Приборостроения",
      74, Ленинград, 1972.

[4]   НОВИКОВ П.С.- Элементы математической логики, Физматгиз,Москва,1959.

[5]   Nilsson N., "A mobile automation: an application of artificial intelligence techniques", Proc. of the 1st
      Int. Conf. on arificial intelligence, Washington, 1969.

[6]   Mc Carthy J., Hayes P.J.,"Some philosophical problems from the standpoint of artificial intelligence".
      Machine Intelligence, 4, Edinburgh, 1969.

[7]   Robinson J.A., "A machine-oriented logic based on the resolution principle". Journal of the Association
      for Computing Machinery, 12, 1, 1965, pp. 23-41.

[8]   Green G., "Theorem-proving by resolution as a basis for question-answering systems". Machine
      Intelligence, 4, Edinburgh, 1969.

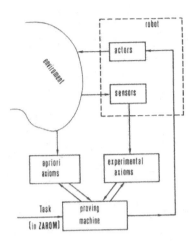

Fig. 1. The diagram of robot control.

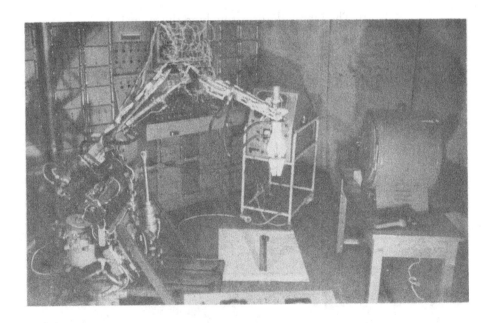

Fig. 2. Our manipulator in work:

a. The visual system is seeing the steam turbine shovel and manipulator is reading.

b. The manipulator is touching the shovel.

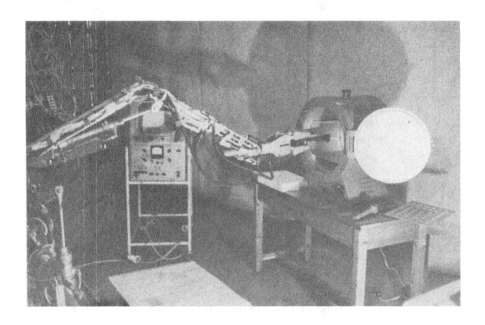

c. The manipulator is placing the shovel in the furnace.

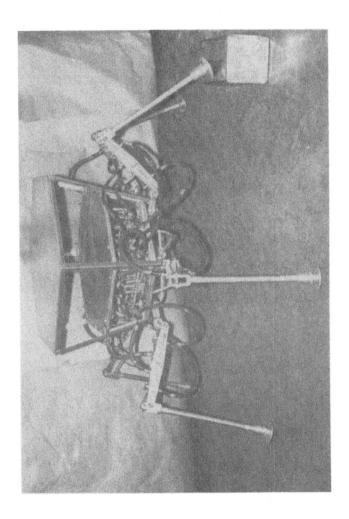

Fig. 3. Our six-leg round machine

ABSTRACT ALGEBRAIC DESCRIPTION METHODS OF A FUNCTIONING PROCESS OF SOME KIND OF ADAPTIVE ROBOTS

M.V. ARTIUSHENKO, Senior Researcher
A.I. KUKHTENKO, Academician
V.N. SEMENOV, Senior Researcher

Institute of Cibernetics, Kiev Ukrainian
Soviet Socialist Republic, USSR

(*)

РЕЗЮМЕ

Проводятся исследования по выбору единово математического аппарата для описания поведения адаптивных роботов.Делается вывод о том, что абстрактно-алгебраический язык описания может найти широкое применение для разработки теоретических основ и инженерных методов проектирования адаптивных роботов.Методами теории групп Ли описываются задачи распознавания образов,ориентации в пространстве, принятия решения, управления движением робота и его исполнительных органов.

(*) All figures quoted in the text are at the end of the lecture

Introduction

In the last years the investigations attached to constructing adaptive type robots [1-4] excited a still greater interest. Therefore it is quite natural to wish to realize, on which mathematical base it is most expidient to describe adaptive robot behaviour. Such a question arises just by reason of the fact that the adaptive robot functioning theory must cover a whole group of problems considered earlier in most different branches of knowledge. First of all we may name : 1) the pattern recognition problem : 2) the space orientation problem ; 3) the decision-making problem (artificial intelligence problem) ; 4) the problem of controlling the movement of the robot and that of its actuators ; and a number of other problems related to elaboration of theoretical principles and engineering approaches of projecting the robot as a very specific machine.

If one is concerned only with the above-mentioned four problems, it may be shown that an abstract algebra apparatus places at an investigator's disposal large possibilities sufficient for studying these, one would think, absolutely heterogeneous questions on the basis of a common mathematical platform. The present paper is devoted to an actual demonstration of some of these possibilities.

Many problems of studying adaptive type robot behaviour may be (with a certain degree of idealization) reduced structurally to the scheme represented in Fig. 1, which is typical for adaptive systems of automatic control, which have also pattern recognition devices, learning devices, self-adjusting devices, logical blocks, decision-making units and others. The standpoint with regard to the expedience of a united theoretical elaboration permitting the study of heterogeneous problems (of pattern recognition, learning, identification, control, etc.) for adaptive control system theory is not new. Professor Y. Z. TSYPKIN's book [5], in which all these diverse problems were considered earlier in a separate way on the basis of a probabilistic iterative process theory (of stochastic approximations), convinces us of the possibility in principle and the attainability of this purpose.

In the present paper abstract algebra, in general and the Lie group theory and the Lie algebra theory, in particular, are used. Our aim remains the same one as in [5], i.e., to demonstrate the sufficiently universal character of the mathematical apparatus, chosen in this case for solving problems, with which one meets while constructing the adaptive robots behaviour theory.

Description possibilities in abstract algebraic terms of the problems indicated in the beginning of the paper are based on the following considerations [6]. The classic mathematical analysis methods, applied usually while describing and studying dynamics problems are introduced into an abstract algebraic theory (and, in particular into the group theory) in general in two ways : 1) through infinitesimal transformations of fundamental group of the dynamic system representation space or 2) through components of an infinitesimal displacement of the local coordinate system (a moving basis).

If one considers a representation space of the system S^n as a union of the connected manifold S and of the transitive group G, which acts in S, and if one denotes through x_j the old variables and through x_i the new variables, then

$$x_i' = f^i(x_j, a^\alpha); \quad i, j = 1, \ldots, n; \ \alpha = 1, \ldots r \tag{1}$$

are equations determining G as a fundamental group, and G is called a Lie group if the funcions f^i are differentiable, both with respect to the variables x_j and with respect to the variables a^α (which are coordinates of a point in the parameter space A of the group G), the number of desired times.

If we take the parameter values $a^\alpha + da^\alpha$, then we shall obtain a transformation T_e transferring the point $W(a^\alpha)$ in $W'(a^\alpha + da^\alpha)$, W, W' $\in$ A. In this way an infinitesimal transformation $T_e = T_a^{-1} T_{a+da}$ is obtained, which relates to a moving coordinate system (a basis). If we denote through x a point with coordinates x_i, then we shall have

$$x' = T_a^{-1} T_{a+da} x, \quad \text{or} \quad T_a x' = T_{a+da} x, \tag{2}$$

whence, when taking into account (1), it follows

$$f^i(x + \delta x; a) = f^i(x; a + da). \tag{3}$$

If we limit infinitesimals in this quality to first order, then after the Taylor's formula, we obtain

$$\frac{\partial f^i}{\partial x_j} \delta x_j = \frac{\partial f^i}{\partial a^\alpha} da^\alpha, \tag{4}$$

which may be represented as

$$\delta x_i = \xi_i^\alpha(x) \ W_\beta^\alpha(a) da^\beta = \xi_i^\alpha(x) \ \omega^\alpha(a, da), \tag{5}$$

where ω^α are linear forms of parameter differentials, i.e. Pfaff's forms and $\xi_i^\alpha(x)$ are functions of a class C^k, $\alpha, \beta = 1, \ldots, r$.

We have for an arbitrary scalar function $F(x)$

$$(6) \qquad \delta F = \frac{\partial F}{\partial x_i} \, \delta x_i = \frac{\partial F}{\partial x_i} \, \xi_i^\alpha(x) \, \omega^\alpha(a, \, da),$$

where $\partial F/\partial x_i \;\; \xi_i^\alpha(x)$ are infinitesimal operators of the group G, and the Pfaff forms ω^α are relative components of the basis (repère).

When studying infinitesimal operators $L = \partial/\partial x_i \;\; \xi_i^\alpha(x)$ Sophus Lie created the operator theory named after him, and Eli Kartan, while studying relative components of the basis, has investigated the more difficult problem attached to a structure, a topology, and a general geometry of the Lie groups. According to [6], the infinitesimal operators and the relative components are, as a matter of fact, two sides of the same formalism of the object, which may be taken as a principle of the basis of an abstract algebraic approach to the description and investigation of different problems.

In § 1 the problems of recognition and orientation are considered, on the whole, from the point of view of the first approach (S.Lie), and in § 2 the control problems are studied in accordance with the second approach (E. Kartan).

Before passing to the actual aspect of the matter, let's note that such a united approach to heterogeneous problems is highly necessary, and is particularly expedient in the cases when one studies functioning fugacious processes, when a separate study, for example, of system logic actions and of a control process dynamics is impossible. In particular : a) problems of investigating the dynamic stability of diverse locomotional system behaviour [3, 7], if the functioning of the upper levels of a hierarchical control system is described in the systems of such a kind by some logic correlations ; b) problems of coordinating the control of the robot's transporting part and its actuators with a sufficiently fugacious processes of functioning, belong here.

Let's make one more general remark in regard to other possible methods of studying the dynamic behaviour of adaptive robots.

Here we mean that other methods originated in the analytic mechanics must also be used, as we imagine it, for the last aim. One of the authors already paid attention to professor Y.I. Grdina's early studies (1911-1913) on living organism dynamics and to the expedience of using these works for studying devices imitating a living organism's behaviour [8, 10]. When applied to the problem of studying adaptive robot behaviour dynamics, these investigations performed by professor Y.I. Grdina at the beginning of our century acquire a particular interest and an unquestionable actuality. The stating of the possibilities revealed here requires however an independent paper. It is just as interesting and expedient to use the other investigations also performed within the analitic mechanics. The question concerns nonholonomic system dynamics studies. One may reveal that a number of locomotional systems and in general devices designed for transferring robots are systems with nonholomic relations. With this, a particular interest is represented by the circumstance that lead to special conditions [11] for nonholonomic dynamic systems while studying motion stability questions even for stationary regimes. The above-mentioned circumstances consist of the fact that the nonholonomic dynamic systems of similar form do not possess one isolated equilibrium state, but there is a whole manifold of equilibrium states, and this attaches particular requirements to the methods of studying the stability of systems of such a kind. The stating of these questions in more detail requires also a special report.

1. A recognition problem and a visual space orientation problem

The problems of : 1) pattern recognition, and 2) visual space orientation, differing by their contents are connected into a united whole for the robot. Possible ways of solving these problems from the position abstract algebra will be indicated below. If one denotes through G a set of robot transferences in a space, then G turns out to be a group of movements with six essential parameters which may be considered as coordinates characterizing the robot movement as a solid,

$$g_i = \{x_{oi}, y_{oi}, z_{oi} ; \psi_i, \vartheta_i, \gamma_i\}, \quad g \epsilon G$$

One can put in correspondence to each element g_i of the set G a determinate perspective image of the visible space. Thus a robot position set generates an image set N. In the general case they are continuum sets and the correspondence $N \leftrightarrow G$ is one-to-one.

The robot "brain" can solve the visual space orientation problem, if one puts into it an effective procedure of establishing the correspondence of the type $N_i \leftrightarrow g_i$, where $n_i \epsilon N$. The static fragment set of a dynamic visual situation perceived by the robot may be split into a finite number of classes N_j by the sign "likeness", with $N = \bigcup_{i=1}^{r} N_j$. Each class N_j has as before the continuum power.

Let's consider primarily the case when n_i is a plane image of a visual situation of plane bodies. With this, each image of the class N_j may be obtained by means of one representative (of the reference image) and of a set of homomorphisms of the image plane f_j^i , where j = 1, 2.

On the other hand, the function set f_j^i may be described by a set of automorphisms P of the picture plane π and it forms an eight parameters Lie group [6, 12]. Thus one succeeds in obtaining the description of an infinite set of images of the visual situation by means of the reference image and of the projective group P 3 p_i of picture plane transformations. The establishing of the correspondence between movement special coordinates and object images is reduced to establishing a homomorphism of the transformation groups G $\overset{\varphi}{\to}$ P.

If one denotes through X and Y, respectively, the sets of the movement space coordinates and of the picture plane, and through A and B the parameter spaces of corresponding transformation groups, then group equations may be written in the form

$$X \to f_1(X,A) \; ;$$
$$Y \to f_2(Y,B) \; ,$$

and one succeeds in determining, by simple reasonings, a homomorphism φ' of parametric spaces of these transformation groups

(7) $A = \varphi'(B) .$

The parameter space B of the projective transformation group P may be given by means of the group's moving basis. With such a consideration, when knowing the reference position of the four characteristic points of the image (the three of which do not belong to the same line) forming the basis R_0 and a new position of the points corresponding to them which give the basis R_1, one determines the point b_i of the parametric space B 3 $b_i = \{ b_1, b_2, \ldots, b_8 \}$.

In following expression (7) we determine the point desired of the parametric space A of the movement group G, and we find thereby spacial coordinates of the robot movement in the basis chosen. When making similar transformations in time there always exists the possibility of determining the first and the second derivatives of the movement's spacial coordinates with respect to time.

The second approach to the visual space orientation problem is based on the possibility of determining the kind of manoeuvre performed by the robot during a global transference by means of tracking movement trajectories of the characteristic points on the two-dimensional manifold formed while considering the picture plane as a projective one and called (later on) the visual manifold.

With this in view let's map the movement group G into the projective transformation group P (determined on the visual manifold M) by means of the homomorphism $\varphi : G \to P$. A transformation set Q of the visual manifold M depending on movement group parameters G arises with this. In considering special cases of a global movement in the space we form one-parametric Lie groups :

$$Q_{x_0}, Q_{y_0}, Q_{z_0}, Q_{\vartheta_0}, Q_{\gamma_0}, Q_{\psi_0} . \tag{8}$$

An element orbit $x \in M$ of the visual manifold :

$$\Omega(x) = \{ \tau_q(x) : q \in Q_i \}, \quad i = x_0, y_0, z_0, \vartheta, \gamma, \psi,$$

where τ determines an action of the group Q_i on the set M, $\tau : Q \times M \to M$ in accordance with the formula

$$(q,x) = \tau_q(x) = x' ; \quad x' \in M, \quad q \in Q_i$$

may be determined for each of such one-parametric groups. It is must convenient to pass to considering the determined group Q_i by introducing for each member an infinitesimal operator

$$L = X_1(x,y)\frac{\partial}{\partial x} + X_2(x,y)\frac{\partial}{\partial y} , \tag{9}$$

whose components X_1 and X_2 are determined as derivatives with respect to a corresponding parameter t of functions ϕ forming a transformation group and satisfying the correlations :

$$X_1 = \frac{d\phi_1}{dt}\bigg|_{t=0} ; \quad X_2 = \frac{d\phi_2}{dt}\bigg|_{t=0} .$$

A trajectory described by a characteristic point on the manifold M is an orbit of the corresponding one-parametric group and it may be found by solving the equation

$$\frac{dx}{X_1(x,y)} = \frac{dy}{X_2(x,y)} .$$

We consider a subset N of M as invariant with respect to the action of some group of transformations Q, if $\forall q \ \tau_q \ (N) \subset N$. It is known that transformation group orbits are invariant curves of the group. By virtue of this and when taking into consideration the local invariance condition

(10) $LF\ (x,y) \equiv 0,$

where $F(x,y) = C$ is a trajectory equation, C = const, we obtain an effective procedure for determining the type of robot movement in a space which consists of the following. During some movement in a space the characteristic points, being patterns of points of a space surrounding the robot, describe on M the trajectories given by an equation $F(x,y) = C$. With this condition, local invariance (10) with respect to the operator (9) giving one of the groups (8), will be fulfilled. The finding of such an infinitesimal operator from the six possible ones L_i determines the above form of the manoeuvre. The method considered is extended to the sufficiently general cases of spacial orientation by means of constructing the corresponding Lie algebra.

When enlarging the system, infinitesimal operators giving groups (8) with two L_p and L_{yp} added, an eight-dimensional vectorial space is formed for which the operators

(11) $L_{xo}, L_{yo}, L_{zo}, L_{\vartheta}, L_{\gamma}, L_{\rho}, L_{yp}, L_{\psi}$

make a basis. One may compose of these operators linear combinations. The operations of finding commutator [6] :

$$[L_{\rho}, L_{\sigma}] = c_{\rho\sigma}^{k} L_{k}$$

are given by a product in this vectorial space of quantities $\sum_{\rho} C_{\rho} L_{\rho}$. Moreover the vectorial space considered is closed with respect to the product introduced which generates a Lie algebra structure in this space.

The Lie algebra considered makes it possible to reduce the problem of determining the form of a manoeuvre performed by the robot to finding a linear combination of infinitesimal operators (11) with respect to which the trajectories of characteristic point movements on the visual manifold would be invariant curves. The local invariance condition will be written in this case in the form

(12) $ALF(x,y) \equiv 0,$ where $A = \sum_i C_i L_i$, C_i = const.

The structural constants C_i carry manoeuvre quantitative characteristics, and the infinitesimal operators taken part in forming linear combination (12), qualitative characteristics.

Let's consider schematically a volumetric body recognition problem.

Unlike the considerations carried out earlier, let's suppose that now some volumetric body V given by two optical mappings of its surfaces A_v^1 and A_v^2 on the visual manifold is to be recognised. One of the approaches to solving volumetric body recognition problems is based on the possibility of determining a form of the transformation P_j destroying an existent discrepancy of these two images for all corresponding points. Thereby a correspondence is established between the points A_v^1 and A_v^2. The form of transformation P_j contains information about the elevation of each point of the body in question V over the horizontal plane. Now the two-dimensional visual manifold M complemented with such a characteristic is reduced into a one-to-one correspondence with two-dimensional space K of the body $V : M \xrightarrow{f} K$. We determine the coordinates of visual points of an object surface in performing the transformation f . For the problems solved by adaptive robots of many kinds one deals most often with recognition cases requiring a reception of object invariant descriptions with respect to translations, rotations, movement, size, form. By means of corresponding three-dimensional groups of the transformations

$$(V_{ref})G \text{ translations;} \quad (V_{ref})G \text{ rotations;} \quad (V_{ref})G \text{ movements;}$$

$$(V_{ref})G \text{ of size;} \quad (V_{ref})G \text{ affine;}$$

Thus, one may judge the membership of the object to the class given in studying the corresponding parametric spaces of three-dimensional groups.

Some other formalism, but of an abstract algebraic character as a matter of fact, may be used when describing and posing decision-making problems. The corresponding possibilities are elucidated in part, for example, in work [13].

2. A control problem

When studying robot functioning processes one comes in a natural way to the notion of hierarchic structure of their control system [4, 7]. If the lower level is described usually by a differential equation language, the upper levels require the utilization of the mathematical logic apparatus and even of description linguistic methods. As M. Arbib and R. Kalman have indicated [14], the description of logic discrete devices (for example, of finite automata) and of ordinary linear dynamic control systems may be made by means of the same abstract algebraic apparatus, — of modulus theory language over a polynomial ring. In [15] a possibility of using the modulus theory language over the differential form ring for studying non-linear dynamic systems is shown.

Below the possibility of using analogous abstract algebraic means for describing the control problem will be demonstrated and their relation with the apparatus used in § 1 for solving the problems of recognition and orientation will be considered.

Let the space of representing a dynamic system (for example, its phase space) belong to some compact open domain of the manifold differentiated M. If $\eta \in$ M is a manifold point, $X_\eta \in TM_\eta$ is a tangential vector, then the mapping X : $\eta \to X_\eta$ determines on M a vectorial field which also may be considered as linear mappings of a set FM of real functions onto M into an algebra of real valued functions on M, i.e. $X(f,g) = (Xf)g + f(Xg)$, where $Xf(\eta) = X_\eta f$; $f,g \in$ FM, $\eta \in$ M.

It turns out that a set X(M) of all vectorial fields on M possesses the properties :

1) X(M) is a vectorial space over a real field R, as

$$(X + Y)f = Xf + Yf, \quad (\alpha X)f = \alpha(Xf), \quad f \in FM, \quad \alpha \in R;$$

2) X(M) is FM-modulus : $(fX)g = f(Xg)$; $f,g \in$ FM

3) X(M) is a Lie algebra over R, i.e. a bilinear operation (product) $[X,Y] = -[Y,X]$ is determined on X(M) such that

$$([[X,Y],Z] + [[Y,X],Z] + [[Z,X],Y]) = 0 .$$

A mapping derived φ' of the space X(M) onto itself corresponds to the afore-mentioned properties of vectorial fields :

1) φ' is a linear transformation;

2) $\varphi'(fX) = (\varphi f)(\varphi' X)$, $f = FM$, $X \in X(M)$;

3) φ' is an automorphism of the Lie algebra X(M):

$$[\varphi' X, \varphi' Y] = \varphi'[X,Y] .$$

If the set of transformations $\varphi_t (-\infty < t < +\infty)$ possesses the properties:

1) φ_t is a diffeomorphism,

2) $\varphi : (t,\eta) \to \varphi_t(\eta)$,

3) $\varphi_{t+s} = \varphi_t \cdot \varphi_s$,

then φ_t forms a one-parameter group of diffeomorphisms of the manifold M which induces a vectorial field X.

Moreover, if M is a smooth (of the class C^k, $K \geqslant 2$) manifold, F : M $\to$ TM is a vectorial field, and a vector F(x) ϵ TM is different from the zero vector TM_x ϵ TM only in the compact part of the manifold M, then there exists a one-parameter group of differeomorphisms φ_t : M $\to$ M, for which the field F(x) is a field of phase velocity :

$$\frac{d}{dt} (\varphi_t x) = F (\varphi_t x) \tag{13}$$

With this any solution of differential equation (13) written usually in the form $\dot{x} = F(x)$, x ϵ M, may be continued unlimitedly.

Thereby it is shown that we can pass from an abstract algebraic description of the dynamic system to describing it in terms of differential equations.

It can also be shown that the vectors $(\partial/\partial x_i) \epsilon TM_x$, i = 1, ..., n, form a basis of the vectorial field F(x, u(t)) linear on u(t) may be represented as a submodulus of FM-modulus X(M), i.e. it may be described in the form [15] :

$$F(x, u(t)) = \frac{dx}{dt} = \sum_{i=1}^{n} \left(g_i \frac{\partial}{\partial x_i} \right) x_i + \sum_{j=1}^{r} \left(h_i \frac{\partial}{\partial u_j} \right) u_j . \tag{14}$$

The set of differentiations $\{\partial/\partial x_i\}$ exactly determines the structure of vectorial fields as a structure of infinitesimal operators of the Lie group considered above, and the set $\{a/au_j\}$ determines so called achievable generatrices of the dynamic system of the FM-modulus [14, 15].

Further, in introducing by means of a conjugated differential operator or by constructing a dual basis dx such that $(\partial/\partial x_i, dx) = \delta_i^j$ is Kronecker's delta, FM*-modulus conjugated with the FM-modulus of vectorial fields, one may obtain a representation of the dynamic system as a submodulus of the modulus of linear differential forms in the shape :

$$\omega_{F(xu)} = \omega_g + \omega_h = \sum_{i=1}^{n} g_i (x) dx_i + \sum_{j=1}^{r} h_j (x) du_j \tag{15}$$

and it is possible to study the dynamic system this time by means of Eli Kartan's formalism, its relation with Sophus Lie's theory has been given attention above.

These ideas stated in a sketshy way should be considered only as the preliminary ground of the principle and possibility of using the abstract algebraic approach to solving the problems posed above. Of course, the solution of each of these problems requires a detailed elaboration and it cannot be elucidated in detail in this short report.

REFERENCES

[1]   KATYS G.P., MAMIKONOV Y.D. et al., "Informational robots and manipulators„ Publishing House, "Energy", Moscow-Leningrad, 1968.

[2]   KOBRINSKII, A.Y.,"Here they are the robots„ Publishing House "Science", Moscow- Leningrad, 1972.

[3]   VUKOBRATOVICH M., "Artificial locomotional systems : dynamics, control algorithms, stability„ (Author's essay of the doctor thesis, Moscow, 1972).

[4]   IGNATYEV M.V., KULAKOV F.M., POKROVSKY A.M., "Algorithms for controlling robots-manipulators„ Publishing House "Machine-building", Leningrad, 1968.

[5]   TSYPKIN Y.Z., "Adaptation and learning in automatic systems„ Publishing House "Science", Moscow-Leningrad, 1968.

[6]   KARTAN E., "The finite continuous group theory and the differential geometry stated through the movible basis„ Publishing House of the Moscow State University, Moscow, 1963.

[7]   OKHOTSIMSKY D.Y., PLATONOV A.K. et al., "Modelling a stepping apparatus on digital computers„ Proceedings of the Academy of Sciences of the USSR, Technical Cybernetics, Issue N.3, 1972.

[8]   GRDINA Y.I.,"Living organism dynamics„ Yekaterinoslav, 1911.

[9]   GRDINA Y.I., "A complement to the living organism dynamics„ Yekaterinoslav, 1913.

[10] KUKHTENKO A.I.,"On a dynamics of devices simulating living organisms„ Transactions of the First Congress of IFAC, vol. II. A theory of discrete optimal and self-adjusting systems, Publishing House of the Academy of Sciences of the USSR, Moscow, 1961.

[11] NEUMARK Y.I. and FUFAYEV N.L.,"A dynamic of nonholonome systems„ Publishing House "Science", Moscow-Leningrad, 1967.

[12] ARTIUSHENKO M.B.,"A visual piloting Lie's algebra„ Coll. "Cybernetics and computing engineering„ N.19, Publishing House "The scientific thought", Kiev, 1973.

[13] BENERGI R., "A theory for solving problems„ Publishing House "The world", Moscow, 1972.

[14] KALMAN R., FALB M., ARBIB M.,"Mathematical system theory essays„ Publishing House, "The world", Moscow, 1971.

[15] KUKHTENKO A. I., SEMENOV V.N., UDILOV V.V.,"An abstract system theory. A contemporary state and development tendences„ In coll. "Cybernetics and computing engineering", Issue N.15, Publishing House "The scientific thought", Kiev, 1972.

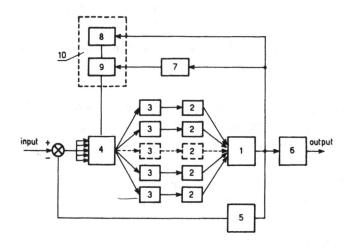

1 – object of control
2 – servomotors of controllers
3 – converters and amplifiers of controllers
4 – decision making block
5 – elements of feedback channel
6 – observing device
7 – teaching device
8 – marks distinguishing circuit
9 – recognizing device classificator
10 – device of recognition

Fig. 1

COMPUTER MANIPULATOR CONTROL, VISUAL FEEDBACK
AND RELATED PROBLEMS

Aharon GILL, Richard PAUL, Victor SCHEINMAN

(*)

Summary

The design of a six degree of freedom manipulator with high positional accuracy and approximately human reach and motion properties is discussed.

The manipulator is moved along time-coordinated space trajectories in which velocity and acceleration are controlled. In lifting and placing objects an approach normal to the support surface is used.

The manipulator is servoed by a small computer which has as input the joint positions and velocities together with other sensor signals. The outputs of the control computer are the joint motor drive levels. No analog servo is used. The servo is compensated for gravity loading and for configuration-dependent dynamic properties of the manipulator.

A set of programs for precise manipulation of simple planar bounded objects by means of visual feedback have also been developed. The image of the hand and manipulated objects is acquired by the computer through the camera. The stored image is analyzed using a corner and line finding program developed for this purpose. The precision obtained is better than .1 inch. It is limited by the resolution of the imaging system and of the arm position measuring system.

(*) All figures quoted in the text are at the end of the lecture

The authors wish to thank Professor Jerome Feldman for his invaluable help and advice in relation to this work.

This research was supported in part by the Advanced Research Projects Agency of the Office of Defense under Contract No. SD-183.

The views and conclusions in this document are those of the authors and should not be interpreted as necessarily representing the official policies, either expressed or implied, of the Advanced Research Projects Agency or the U.S. Government.

1. Introduction

This paper describes work at the Stanford Artificial Intelligence Project on Manipulation and Visual feedback. The work is described extensively in [Gill], [Paul] and [Scheinman] and this paper is an attempt to provide an overview of the combined work. We first describe the design of the manipulator shown in Figure 1 and then go on to describe in detail the trajectory generation for manipulator motions and the software servo system. Final sections describe the corner finder used by the visual feedback system and the visual feedback tasks.

2. Arm Design

The arm is a manipulator with six degrees of freedom and is characterized by two rotating "shoulder" joints, a linear motion "elbow" joint and three rotating "wrist" joints. A vise grip hand serves as the terminal device (see Figure 1).

The results of systems studies indicated that an electric motor powered, "Unimate" type arm was best suited to our tasks and goals. By making the axes of joints 4, 5, and 6 intersect at one point, this arm is a "solvable configuration" [Piper]. Furthermore, to facilitate computation, all axes are made either normal or parallel to adjacent axes.

The two shoulder joints are powered by printed circuit motors having low moments of inertia. A harmonic drive component set reduces the motor speed by 100/1 and gives torque multiplication of about 90 for both shoulder joints. An electromagneitc brake, mounted on the high speed motor shaft, holds the joint at any given position, eliminating the need for continuous application of motor torque.

The two shoulder joints are constructed of large diameter tubular aluminum shafts which house the harmonic drive and slip clutch. Each tubular shaft is mounted on two large diameter, thin section ball bearings supported in a solid external housing. As these two joints act on small moment arms, heavy sections are liberally employed because inertia affects of added mass are relatively small in this area of the arm. Each of these two joints has integral, conductive plastic, potentiometers. The geometrical configuration of joint #1 is such that the arm base can be bolted to any flat surface, but the design calculations have been made with view towards a conventional table mounting.

To allow for maximum useful motion, the elbow joint is offset from the intersection of the axes of joints #1 and #2. This extra link added to the arm geometry, allows 360 degrees of rotation of both shoulder joints.

Joint # 3, the linear motion elbow joint, has a total travel of 24 inches, making the second link length variable from 6 inches to 30 inches. The main boom is a 2.5 inch square aluminum tube which rolls on a set of sixteen phenolic surfaced rollers. These rollers resist tube twisting and bending moments and support normal loads. They allow only pure, one-dimensional linear motion of the tube. The sections of the square tube boom and the supporting rollers have been designed to optimize performance with respect to structure deflection, natural frequency, and load carrying ability.

The use of rollers provides a larger bearing surface area than ball slides, etc., while the square, thin-walled tube provides better roller support near its edges than a comparable section of round tube. The drive for this joint is provided by a gear driven double rack mounted along the neutral axis of the outside of the boom. Although inefficient, a worm reducer stage is employed because of the necessary reduction of 10/1 and the desire to minimize the number of components and special tolerances in the gear box. A strip of conductive plastic material is cemented along the center line on the side opposite the boom. This is read by a wiper element mounted inside the roller housing to give a positive indication of the boom position. The # 3 joint drive is a permanent magnet motor, with a brake and tachometer integral in its case.

Design of the wrist joints #4 and #5 is similar to that of the shoulder joints except that all components are smaller and lighter. Great attention has been paid to obtain the required performance with the least mass. Small size harmonic drive components are used with high speed, permanent magnet motors. A shaft brake is also employed to hold position. These motors have 26 bar armatures and exhibit nearly cog free operation.

Joint # 6, the hand rotation, employs a standard 1-1/8 inch diameter, planetary gear reduction and a permanent magnet motor. This drive has about 1 degree backlash at the output shaft.

The last basic component designed is a vise grip hand which employs the sliding finger concept. Here, two interchangeable plate like fingers slide together symmetrically, guided on four rails (two for each finger). They are driven by a rack and pinion arrangement utilizing one center gear and a rack for each finger. The maximum jaw opening is 4 inches with a holding force of 20 lbs.

Thick, rubber jaw pads provide a high coefficient of friction for positive handling of most objects. Each finger pad is provided with a switch type touch sensor mounted in the center of the grip area.

3. Trajectories

In order to move the arm we first calculate a trajectory. This is in the form of a sequence of polynomials, expressing joint angles as a function of time, one for each joint. When the arm starts to move, it is normally working with respect to some surface, for instance, picking up an object from a table. The initial motion of the hand should be directly away from the surface. We specify a position on a normal to the surface out from the initial position, and require that the hand pass through this position. By specifying the time required to reach this position, we control the speed at which the object is lifted.

For such an initial move, the differential change of joint angles is calculated for a move of 3 inches in the direction of the outward pointing normal. A time to reach this position based on a low arm force is then calculated. The same set of requirements exists in the case of the final position. Here we wish once again to approach the surface in the direction of the normal, this time passing down through a letdown point.

This gives us four positions: initial, liftoff, letdown, and final. If we were to servo the arm from one position to the next we would not collide with the support. We would, however, like the arm to start and end its motion with zero velocity and acceleration. Further, there is no need to stop the arm at all the intermediate positions. We require only that the joints of the arm pass through the trajectory points corresponding to these intermediate positions at the same time.

The time for the arm to move through each trajectory segment is calculated as follows: For the initial and final segments the time is based on the rate of approach of the hand to the surface and is some fixed constant. The time necessary for each joint to move through its mid-trajectory segment is estimated, based on a maximum joint velocity and acceleration. The maximum of these time is then used for all the joints to move through the mid trajectory segment.

We could determine a polynomial for each joint which passes through all the points and has zero initial and final velocity and acceleration. As there are four points and four velocity and acceleration constraints we would need a seventh order polynomial. Although such polynomials would satisfy our conditions, they often have extrema between the initial and final points which must be evaluated to check that the joint has not exceeded its working range.

As the extrema are difficult to evaluate for high order polynomials, we use a different approach. We specify three polynomials for each joint, one for the trajectory from the initial point to the liftoff point, a second from the liftoff to the setdown point, and a third from the setdown to the final point. We specify that velocity and acceleration should be zero at the initial and final points and continuous at the intermediate points. This sequence of polynomials satisfies our conditions for a trajectory and has extrema which are easily evaluated.

If a joint exceeds its working range at an extremum, then the trajectory segment in which it occurs is split in two, a new intermediate point equal to the joint range limit is specified at the break, and the trajectory recalculated.

A collision avoider will modify the arm trajectory in the same manner by specifying additional intermediate points. If a potential collision were detected, some additional points would be specified for one or more joints to pass through in order to avoid the collision.

4. Servo

The servo program which moves the arm is a conventional sampled data servo executed by the computer with the following modifications. Certain control constants the loop gain, predicted gravity and external torques are precalculated and varied with arm configuration.

We treat the system as continuous, and ignore the effects of sampling, assuming that the sampling period is much less than the response time of the arm. Time is normalized to the sampling period, which has the effect of scaling the link inertia up by the square of the sampling frequency. The Laplace transform is used throughout.

The set point for each joint of the arm is obtained by evaluating the appropriate trajectory segment polynomial for the required time. The velocity and acceleration are evaluated as the first and second derivatives of the polynomials.

The position error is the observed position ϑ less the required value ϑs. Likewise the velocity error is the observed velocity less the required velocity. Position feedback is applied to decrease position error and velocity feedback is used to provide damping.

A simple feedback loop is shown in Figure 2. The arm is represented by $1/s^2 J$, where J is the effective link inertia, a function of arm configuration. $T(s)$ is an external disturbing torque. The set point $R(s)$ is subtracted from the current position to obtain the position error $E(s)$ and is multiplied by s, representing differentiation, to obtain the error velocity. There are two feedback gains ke and kv, position and velocity respectively.

By writing the loop equation we can obtain the system response:

(1) $E(s) = (-s^2 J)/(s^2 J + skv + ke) * R(s) + 1/(s^2 J + skv + ke) * T(s)$

and the condition for critical damping is:

(2) $kv = 2(J * ke)^{1/2}$

It can be seen that the system response is dependent on J as would be expected. Because the effective link inertia J can vary by 10:1 as the arm configuration changes, we are unable to maintain a given response (see Equation 2) independent of arm configuration. If, however, we add a gain of $-J$ as shown in Figure 3 then we obtain:

(3) $E(s) = (-s^2)/(s^2 + skv + ke) * R(s) + 1/(s^2 + skv + ke) * T(s)/J$

and the condition for critical damping is:

(4) $kv = 2 * (ke)^{1/2}$

It can be seen that the servo response is now independent of arm configuration.

The principal disturbing torque is that due to gravity, causing a large position error, especially in the case of joint 2. If we were able to add a term equal to the negative of the gravity loading Tg (see Figure 3) then we would obtain the same system response as in Equation 3 except that T would become Te, the external disturbing torque, less the gravity dependent torque, reducing the position error.

We can compensate for the effect of acceleration of the set point R(s), the first term in Equation 3, if we add a term $s^2 R(s)$ (see Figure 3) to obtain finally a system response:

$$E(s) = 1/(s^2 + skv + ke) * T(s)/J$$

The gain of $-J$ and the torque Tg are obtained by evaluating the coefficients of the equations of motion [Paul] at intervals along the trajectory.

The servo has uniform system response under varying arm configurations and is compensated for gravity loading and for the acceleration of the set point r.

Although these gains give an acceptable response from the point of view of stiffness, the gain is too low to maintain the high positional tolerance of -0.05 inches, which we are just able to measure using the 12 bit A/D converter. In order to achieve this error tolerance, the position error is integrated when the arm has reached the end of its trajectory. When the position error of a joint is within tolerance the brake for that joint is applied and the joint is no longer servoed. When all the joints are within the error tolerance the trajectory has been executed.

The output of the servo equation is a torque to be applied at the joint. Each joint motor is driven by a pulse-width modulated voltage signal. The output of the computer is this pulse-width and the polarity. The drive module relates torque to drive voltage pulse-width.

The motors are driven by a 360 Hertz pulse-width modulated voltage source. The program output "h" is the relative "on" time of this signal. If we plot an experimental curve of "h" v joint torque we obtain two discontinuous curves depending on the joint velocity (see Figure 4).

This curve can be explained in terms of two friction effects: load dependent, causing the two curves to diverge, and load independent, causing separation at the two curves at the origin. The electrical motor time constant also affects the shape of the curve near the origin. Experimentally determined curves are supplied to the servo program in piecewise linear form.

One other factor considered is the back emf of the motor. The value of "h" is the ratio of required voltage to supply voltage. The supply voltage is simply augmented by the computed emf before "h" is calculated.

5. Control

Two programs exist, one for planning "arm programs" and the other for executing the resulting trajectory files. This secion lists the arm primitives, which have meaning at two times: once at planning, when the trajectory file is being created and feasibility must be checked, trajectories calculated etc., and once at execution time when the primitives are executed in the same way that instructions are executed in a computer.

> OPEN (DIST) Plan to open or close the hand such that the gap between the finger tips is DIST.

> CLOSE (MINIMUM) Plan to close the hand until it stops closing and then check that the gap between the finger tips is greater than MINIMUM. If it is less, then give error 2.

> CENTER (MINIMUM) This is the same as CLOSE except that the hand is closed with the touch sensors enabled. When the first finger touches, the hand is moved along with the fingers,

keeping the touching finger in contact. When the other finger touches, both fingers are driven together as in CLOSE.

CHANGE (DX—DY—DZ, VELOCITY) Plan to move the arm differentially to achieve a change of hand position of vector DX—DY—DZ at a maximum speed of VELOCITY.

PLACE Plan to move the hand vertically down until the hand meets some resitance, that is, the minimum resistance that the arm can reliably detect.

MOVE (T) At planning time check that the position specified by the hand transformation T is clear. Plan to move the hand along a trajectory from its present position to | T |. The hand is moved up through a point LIFTOFF given by LIFTOFF = INITIAL POSITION + DEPART, where DEPART is a global vector initialized to z = 3 inches. Similarly on arrival the hand is moved down through a point SET DOWN given by: SET DOWN = FINAL POSITION + ARRIVE, ARRIVE is also set to z = 3 inches.

PARK Plan a move as in MOVE but to the "park" position.

SEARCH (NORMAL, STEP) Set up for a rectangular box search normal to NORMAL of step size STEP. The search is activated to AOJ instruction.

There are also control primitives which specify how the other primitives are to be carried out.

STOP (FORCE, MOMENT) During the next arm motion stop the arm when the feedback force is greater than the equivalent joint force. If the arm fails to stop for this reason before the end of the motion, generate error 23.

SKIPE (ERROR) If error ERROR occurred during the previous primitive then skip the next primitive.

SKIPN (ERROR) If error ERROR occurred during the previous primitive execute the next primitive otherwise skip the next primitive.

JUMP (LAB) Jump to the primitive whose label is LAB

AOJ (LAB) Restore the cumulative search increment and jump to LAB.

WAIT Stop execution, update the state variables and wait for a proceed command.

TOUCH (MASK) Enable the touch sensors specified by mask for the next primitive.

SAVE Save the differential deviation form the trajectory set point. This can be caused by CHANGE type primitives.

RESTORE Cause the arm to deviate from the trajectory set point at the end of the next motion by the deviation last saved.

With the exception of MOVE, which requires a trajectory file, most functions can be executed directly by prefixing the primitive name by "DO". The planning program plans the action and sends it to the arm servo program to be executed. This does not change the state of the arm servo program if it is in a "wait" state and execution can continue after any number of executed primitives. This method is used by the interactive programs, which will plan a move to bring the hand close to the required place and then plan a "wait". When executed, the hand position will be modified during the wait phase by the interacting program executing a series of "DO" commands. Execution of the preplanned trajectory can then continue by calling "DO–PROCEED".

The arm system has been programmed to provide a set of general block manipulation routines. With these routines it is necessary only to give the name of the block and its desired position and orientation; the program then generates the required moves and hand actions to perform the transformation. These routines were used in conjunction with the vision and strategy systems to solve the "Instant Insanity" puzzle [Feldman]. In the case of manipulation tasks, this system has been employed to screw a nut onto a bolt and to turn a crank. With the development of a corner operator visual feedback tasks could be performed.

6. Corner Finder

We will now describe the corner finder and the visual feedback tasks in which it is used. The purpose of the corner finder is to find lines and corners (which are the main features of planar bounded objects) in a small area of the frame of intensity values read into the computer memory from the vidicon camera. The corner finder utilizes information about the features to the extent given to it; it is not a general scene analyzer (even in the context of planar bounded objects), and although it can be used as part of one it will be uneconomical to do so. The corner finder operates by analyzing part of the area (a window) at a time and moving the analyzed window in a controlled search pattern when needed.

Two main types of scene analyzers using simple intensity information have been developed over the years:

(a) The "gradient follower" type looks for boundaries of regions by analyzing intensity gradients at image points.

(b) The "region grower" type aggregates points based on some similarity criterion to form regions.

The corner finder uses ideas from both these types. It makes rough checks on the existence of regions in the analyzed area. For this purpose each point within the area is processed simply to form the intensity histogram of the area. It then follows boundaries of regions by using a dissimilarity criterion. No gradient type processing is used so that continuity is not lost at points of weak gradient, sharp corner, etc. The corner finder is described in detail in [Gill].

General scene analyzers do not use any prior information because there is no reason for them to assume the existence of such information. On the other hand the corner finder described here uses prior information down to its lowest levels. The design philosophy is to use and check against prior information at the earliest possible moment. The corner finder can find only simple corners directly. Complex corners, with more than two edges, can then be constructed from simpler corners. Generally, the vertices and edges of simple corners found in the image will not completely coincide even if the simple corners are parts of the same complex corner. Therefore we will merge them to form a complex corner if they are "close" (within some tolerance), and especially if there is some external information which indicates the existence of a complex corner rather than that of several separate corners. The following assumptions guided the development of the corner finder. They are not all necessary conditions for its operation or success. The most important assumption is that some of the properties of the corner (e.g. location, form and orientation, relative inside to outside intensity are known at least approximately. The properties of the object to which this corner belongs are known (e.g. the hand or a specific cube), or because this corner was found before by the same or similar programs.

Not all the properties need be given to the program. The user or a higher level program can give as many of the properties as he/she/it decides to give. Actually the properties are not only "given" to the program, but the user can "demand" a match, within a given tolerance, of these properties and the actual measured properties of the corner found.

Some comments about window size: the window size which is regularly used has a dimension of 18 x 18 raster units. When the 2 inch focal length lens is used it corresponds to a field of view of approximately 1 degree which incidentally is the field of view of the sensitive part of the human eye, the fovea. The fovea however has about 5 times more sensing elements in the same field of view. We should also note the human ability to resolve between pairs of lines that are closer than the distance between the sensing elements. Carrying the above analogy a little farther we can say that moving the window in the frame is similar to the movement of the eye in the head, while moving the camera is similar to rotating the head.

This size of the window was chosen in order to fulfill the assumptions that the window is smaller than the image of the object so that each line or corner intersects the perimeter of the window, but big enough so that we will have enough boundary points to get a good line fit. Also we want the size of the window to be small enough so that the assumption of almost uniform intensity inside the window is justified.

7. Visual feedback Tasks

The purpose of the visual feedback tasks is to increase the precision of the manipulations done in the hand-eye system. The feedback currently does not take into account the dynamic aspects of the manipulation.

The tasks are carried out in the context of the general tasks that the hand-eye system can currently perform, i.e. the recognition and manipulation of simple planar bounded objects. The manipulations that we sought to make more precise with the addition of visual feedback are grasping, placing on the table and stacking. The precision obtained is better that .1 inch. This value should be judged by comparing it with the limitations of the system. The resolution of the imaging system with the 2 inch lens is 1 mrad. which, at an operating range of 30 inches, corresponds to .03 inch. The resolution of the arm position reading (lower bit of the A/D converter reading the first arm joint potentiometer) is also .03 inch but the noise in the arm position reading corresponds to .05 inch. When we tried to achieve precision of .05 inch, the feedback loop was executed a number of times until the errors randomly happened to be below the threshold.

The question of whether the visual feedback, or in response in general, is dynamic or not, is sometimes more semantic than real. What the person asking the question means in this case is, does it seem to be continuous ? The question then is really that of cycle time or sampling period. A cycle time of 20 msec will suffice to fool the human eye so that the response will be called "dynamic." Since the computations needed and the computing power that we now have cause the length of the cycle time to be several seconds, no attempt was made to speed it up by programming tricks, use of machine language, etc. With this cycle time the movement of the arm actually stops before we analyze the situation again, so that we do not have to take into account the dynamic aspect of the error.

In addition to computing power, the vidicon camera also presents some limitations to faster response. The short time memory of the vidicon which helps us (persons) to view the TV monitor, will "Smear" a fast moving object. If the scene is bright enough a shutter can be used. If the vision can be made sufficiently fast and accurate, the control program currently used to run the arm dynamically could be expanded to incorporate visual information.

One of the characteristics that distinguishes our visual feedback scheme is that the analysis of the scene, to detect the errors to be corrected, is done with the hand (which is still holding the object) in the scene. In the grasping task the presence of the hand is inevitable. In other tasks, for example stacking, being able to analyze the scene with the object still grasped helps to correct the positional errors before they become catastrophic (e.g. the stack falls down). Also some time is saved since there is no need to release the object, move the arm away, bring it back and grasp the object again. We pay for this flexibility with increased complexity of the scene analysis.

The difficulty is lessened by the fact that the hand-mark (which is used to identify the hand see Figure 5) has known form and relative intensity which helps to locate it in the image. In the grasping task the ability to recognize the hand is necessary. The task is essentially to position the hand at fixed location and orientation relative to the object to be grasped.

We have found it to our benefit to locate the hand first in the other tasks also. After the hand-mark is found, we use its location to predict more accurately the locations in the image of the edges of the object held by the hand.

Moreover, after the hand-mark has been found, it will not be confused with other edges in the scene. Since we are using only one camera (one view), we cannot measure directly even differences of locations of two neighboring points. Hence the three-dimensional (3-D) information has to be inferred from the two-dimensional (2-D) information available in the image and some external information.

The external information, which is used in the visual feedback tasks is supplied either by the touch sensors on the fingers of by the fact that an object is resting on the table-top or on another object of known dimensions. The touch sensors help us to determine the plane containing the hand mark from the known position of the touched object. The support hypothesis gives us the plane containing the bottom edges of the top object.

Before an object is to be grasped or stacked upon, the arm is positioned above the object and the hand-mark is sought. The hand is positioned high enough above the object so that the corner finder does not confuse the hand-mark with the object. After the hand-mark is found, the difference between the coordinates of the predicted location of the hand-mark and the location where it was actually found is stored. The same is done for the place on the table where an object is going to be placed.

The table is divided into 4-inch squares, (there are 100 squares), and the corrections are stored with the square over which they were found. When we subsequently look for the hand-mark over this part of the table, the stored differences are used to correct the prediction. Since we now have a corrected prediction, the dimension of the window, or the search space used, can be made smaller.

Each time that the hand-mark is found again, the differences between predicted (before correction) and actual locations in the image are also used to update the stored corrections.

To find the hand we look for both corners, since the scene is complicated by the presence of other objects. The camera is centered on the hand-mark. Using the camera and arm models, the locations of the images of the two lower corners of the hand-mark are predicted. Also the form of the image of the corners is computed. The predicted width of the hand-mark in the image is stored.

Using the information computed above, the right side corner is sought first, using the corner finder. If the corner is found, the error between its predicted and actual locations is used to update the prediction of the location of the left side corner which is now sought. If the right corner is not found we look for the left one first.

This algorithm is an example of the use of information about a relation between two features to be found, in addition to information pertaining to each feature alone.

We check that we found the corners belonging to the hand-mark, and not those belonging to a cube which might have very similar form, by comparing the distance between the corners in the image with the stored predicted width.

8. Grasping

The "Grasping" task is to grasp precisely a cube of approximately known position and orientation in order to move it and place it somewhere else, or stack it on another cube. The precision is needed in order not to drop the cube in midtrajectory (which can happen if the cube is grasped too close to an edge), and in order that its position relative to the hand will be known. This information is used in the other two visual feedback tasks. We try to grasp the cube on the mid-line between the faces perpendicular to the fingers, half way above the center of gravity. Note that in one direction (namely perpendicular to the fingers) the hand does its own error correcting. When the hand is positioned over the cube with maximum opening between the fingers (2.4 inches between the tips of the touch sensors) and then closed, the cube will be moved and always end in the same position relative to the hand, independent of the initial position. This motion is sometimes disturbing (e.g. when grasping the top cube of a stack, the movement can cause it to become unstable before it is fully gripped and it will fall off the stack), and hence no use is made of this feature. Instead we correct errors in this direction as well, such that when a cube is grasped it is moved by less than the tolerance of the feedback loop.

The grasping is done in the following steps:

(a) The fingers are fully opened and the hand is moved over the center of the cube so that the fingers are parallel to the cube's sides.

(b) The touch sensors are enabled and the hand is closed slowly (at about 1/4 of the usual speed or about 1 inch/sec) until one of the fingers touches the face of the cube. The touch is light enough so that the cube is not moved. The touch sensors are then disabled.

(c) Using the distance between the tips of the sensors after the closing motion of the fingers is stopped, the equation for the plane containing the hand-mark facing the camera is computed.

(d) The hand-mark is then sought. After the two corners of the hand-mark have been found, the camera transformation is used to compute the corresponding rays. These rays are intersected with the plane found in step (c) to give the coordinates of the corners. To verify that the corners found do belong to the hand-mark, we check that they have approximately the same height, and that the distance between them corresponds to the width of the hand-mark.

(e) Using the information already used in step (c), the position errors of the hand are computed. If the magnitudes of the errors in all three directions are less than a treshold (currently .1 inch), the task is finished, we go to step (f) and then exit. If the errors are larger, the hand is opened and the errors are corrected by changing the position of the arm appropriately. We then go back to step (b) to check errors again.

9. Placing

The placing task is to place the cube precisely at a given location on the table. With very minor modifications, it could be used to place the cube on any relatively large horizontal surface of known height which does not have any reference marks near the location where the cube is to be placed. In this case the support hypothesis is the only external information used. The task is carried out in the following steps:

(a) The cube is grasped and moved to a position above the table where the cube is to be placed.

(b) The cube is placed.

(c) The hand-mark is located in the image.

(d) The camera is centered on the visible bottom edges of the cube.

(e) The locations of mid-points and the orientation of the images of the two visible bottom edges of the cube are computed using the hand transformation and the size of the cube. The predicted location is then corrected by the amounts computed in step (c).

(f) The corner finder is then used to locate the two lines in the image. The two lines are intersected to find the corner location in the image. Using the support hypothesis, the location of the corner is computed and compared with the required location. If the magnitudes of the errors are less than a threshold (.1 inch) in both directions then the task is completed. Otherwise the cube is lifted and the error corrected by changing the position of the arm appropriately. We then go back to step (b) to check the errors again.

10. Stacking

The stacking task is to stack one cube on top of another cube so that the edges of the bottom face of the top cube will be parallel to the edges of the top face of the bottom cube, at offsets specified to the program by the user or the calling module.

The task is carried out in the following steps:

(a) The top cube is grasped.

(b) The camera is centered on the top face of the bottom cube. The midpoints and orientations of the images of the two edges of the top face of the bottom cube, belonging also to the most visible vertical face and other visible vertical face, are computed. The corner finder is used to locate these two lines in the image. The locations and orientations found are then stored. The two lines are intersected and using the known height of the cube, the location of the corner is found. Using the given offsets, the coordinates of the required positions of the corner of the bottom face of the top cube and the center of the top cube are calculated.

(c) The top cube is moved to a location just above the bottom cube, oriented so that the hand-mark is parallel to most visible vertical face of the bottom cube.

(d) The top cube is placed on the bottom cube.

(e) The hand-mark is located in the image and then the two edges of the bottom face of the top cube are located as in steps (c) to (f) of the placing task. In this case, however, the edges of the top face of the bottom cube will also appear in view. A simple algorithm is used with the information computed in step (b) to decide which of the lines are the edges of the bottom face of the top cube. This simple algorithm can be deceived sometimes by the presence of shadows and "doubling" of edges in the image. We could make the algorithm more immune by locating the vertical edges of cubes also if they could be found.

(f) The two edges of the bottom face of the top cube found in the last step are intersected to find the corner location in the image. Using the support hypothesis, the coordinates of the location of the corner are computed and compared with the required location computed in step (b). If the magnitudes of the errors are less than a threshold (.1 inch) in both directions then the task is completed. Otherwise the cube is lifted and the error corrected by changing the position of the arm appropriately. We then go back to step (d) to check the errors again.

Instead of a bottom cube, we can specify to the program a square hole in a bottom object in which the top cube is to be inserted. In this case, when the top cube is placed on the bottom object we have to check how much it was lowered. If it was lowered past some threshold this means that it is already in the hole and can be released. We make sure that the grip of the hand is tight enough so that the cube grasped will not rotate when placed partly above the hole.

The programs described here are presently being expanded to provide a system capable of discrete component assembly tasks.

REFERENCES

[1]   Feldman J. with others, "The use of Vision and Manipulation to Solve the 'Instant Insanity' Puzzle." Second International Joint Conference on Artificial Intelligence, London September 1-3, 1971.

[2]   Gill A. , "Visual Feedback and Related Problems in Computer Controlled Hand-Eye Coordination," Stanford Artificial Intelligence Memo 178, October 1972.

[3]   Paul R.P.C., "Modelling, Trajectory Calculation and Servoing of a computer Controlled Arm," Stanford Artificial Intelligence Memo 177, March 1973.

[4]   Pieper D.L., "The Kinematics of Manipulators Under Computer Control", Stanford Artificial Intelligence Memo 72, October 1968.

[5]   Scheinman V.D., "Design of a Computer Manipulator", Stanford Artificial Intelligence Memo 92, June 1969.

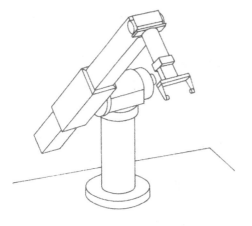

Fig. 1. Arm

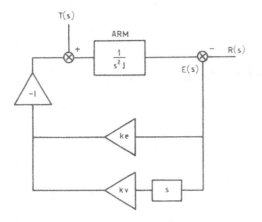

Fig. 2. Simple servo loop.

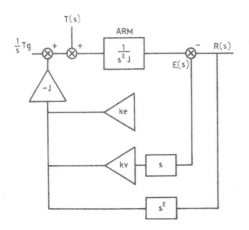

Fig. 3. Compensated servo loop.

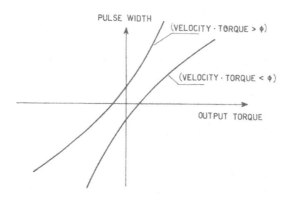

Fig. 4. Pulse width v. Output torque.

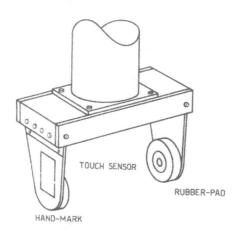

Fig. 5. Hand mark.

MONITORED FINITE STATE MACHINES (*)
(A construction theorem)

Fabrizio LUCCIO, Professore Incaricato,
Istituto di Scienze della Informazione
Università di Pisa,
Pisa, Italy

Summary

   The classical definition of finite state machine has been extended from stationary to time variant systems. Namely, machine structure is defined as a function of time.

   If the structure is to be selected among a finite collection, and such a selection is in turn performed by a finite state machine (the monitor), the total system is equivalent to a proper stationary machine.

   However, the construction of a monitored time variant machine MT equivalent to a given stationary machine Z can be performed, only if special conditions on Z are met.

   A theorem is proved, based on a decomposition of the input alphabet of Z, and on the definition of particular covers on the states of Z, leading to the construction of MT.

(*) The research reported in this paper has been supported in part by the Consiglio Nazionale delle Ricerche, under research grant No. 72.00254.42.

1. Introduction. Stationary and time variant finite state machines

According to a classical approach [1], a finite state machine (or automaton) A can be defined as a formal quartuple

$$(1) \qquad\qquad A = \{X, S, s_1, f\}$$

where:

X is a finite set of inputs that can be applied to A;

S is a finite set of states that can be assumed by A;

$s_1 \in S$ is the initial state of A;

$f: S \cdot X \rightarrow S$ is the next state function, specifying the next state $f(s,x) \in S$ assumed by A for any pair $s \in S, x \in X$.

Model (1) corresponds to a machine A to work in discrete time $(t = 0, 1, 2, \dots)$. Initially A is in state s_1. Then an input sequence

$$(2) \qquad x^* = x_1 x_2 \dots x_n \qquad x_i \in X, \qquad i \in \{1, 2, \dots, n\}$$

is applied to A, where any x_i occurs at time $t = i$ and drives the machine, in a unity of time, from the present state s to the next state $f(s, x_i)$. Hence, the final state reached by A at time $n + 1$, for input sequence x^*, is given by

$$f(\dots f(f(s_1, x_1), x_2) \dots, x_n).$$

Machine (1) has been occasionally termed a state-automaton, in the sense that the output produced by A in response to a given input sequence corresponds to the sequence of states assumed by A. If a different output behavior is to be specified, model (1) can be modified by the addition of a proper set of outputs Y, and of an output function defined on S x X onto Y. The output problem will not be discussed in this paper, since it is not relevant to the machine properties to be investigated here.

The mathematical model considered thus far corresponds to a stationary (i.e. time invariant) machine. The idea of time variant machines has been firstly introduced by Gill [2], and then refined by different authors [3, 4]. Here, we propose the following definition of a time variant finite state machine (briefly t.v.m.) T, that directly derives from the model of [4]:

$$(3) \qquad\qquad T = \{X, S, s_1, f^*\}$$

where

X, S and s_1 have the same meaning of (1);

$f^* = f_1 f_2 f_3 \ldots$ is an infinite sequence, where any f_i is a particular next state function, $f_i : S \times X \to S$, to be applied at time $t = i$.

The set of all possible functions f_i is denoted by F. Note that, since F has finite cardinality, $\#(F) = \#(S)^{\#(S) \cdot \#(X)}$, at least one particular function f_i has infinite occurrences in f^*.

Thus, the application of an input sequence x^* defined as in (2) drives T, at time $n + 1$, to the state

$$f_n (f_{n-1} (\ldots f_2 (f_1 (s_1, x_1), x_2) \ldots, x_{n-1}), x_n).$$

If an additional set C of control inputs is considered, a control function g can be defined on C onto F:

$$g : C \to F$$

to establish a correspondence between control inputs c_j and next state functions f_i. Then, model (3), at time $t = i$, can be visualized as:

$$(4)$$

Generation of the control input sequence is the subject of the next section.

2. Finite state monitors

Let us make the assumption that the control input sequence to a t.v.m. $T = \{ X^T, S^T, s_1^T, f^{*T} \}$ is generated by a stationary finite state machine $M = \{ X^M, S^M, s_1^M, f^M \}$, called monitor. That is, $S^M \equiv C^T$, and the configuration results:

(5)

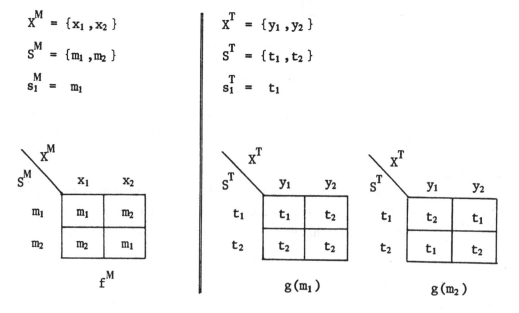

The whole configuration (5) is equivalent to a stationary finite state machine Z defined by:

$$Z = \{X^Z, S^Z, s_1^Z, f^Z\}$$

$$X^Z = X^M \times X^T;$$

(6)

$$S^Z = S^M \times S^T; \qquad s_1^Z = (s_1^M, s_1^T);$$

$$f^Z((s_k^M, s_j^T), (x_a^M, x_b^T)) = (f^M(s_k^M, x_a^M), g(s_k^M)(s_j^T, x_b^T)).$$

For example, assume that machines M and T are defined by:

$$X^M = \{x_1, x_2\} \qquad\qquad X^T = \{y_1, y_2\}$$

$$S^M = \{m_1, m_2\} \qquad\qquad S^T = \{t_1, t_2\}$$

$$s_1^M = m_1 \qquad\qquad\qquad s_1^T = t_1$$

Then, the equivalent machine Z results:

$$X^Z = \{(x_1,y_1),(x_1,y_2),(x_2,y_1),(x_2,y_2)\}$$

$$S^Z = \{(m_1,t_1),(m_1,t_2),(m_2,t_1),(m_2,t_2)\}$$

$$s_1^Z = (m_1,t_1)$$

$f^Z:$

$S^Z \diagdown X^Z$	x_1,y_1	x_1,y_2	x_2,y_1	x_2,y_2
m_1,t_1	m_1,t_1	m_1,t_2	m_2,t_1	m_2,t_2
m_1,t_2	m_1,t_2	m_1,t_2	m_2,t_2	m_2,t_2
m_2,t_1	m_2,t_2	m_2,t_1	m_1,t_2	m_1,t_1
m_2,t_2	m_2,t_1	m_2,t_2	m_1,t_1	m_1,t_2

(7)

Machine Z can always be directly constructed from the specification of M and T. However, the opposite construction seems to be relevant in the design area: that is, the decomposition of a given stationary machine Z into a t.v.m. T, controlled by a monitor M. To present a basic construction theorem, new definitions are needed, some of which pertaining to the classical field of automata decomposition. Namely:

A cover μ on the set of states S^Z is a collection of proper subsets of S^Z (called blocks), $\mu = \{B_1^\mu, B_2^\mu,..., B_m^\mu\}$, $B_i^\mu \subset S^Z$ for any $i \in \{1,2,...,m\}$, such that $\bigcup_{i=1}^{m} B_i^\mu = S^Z$, and $B_i^\mu \not\subseteq B_j^\mu$ for any i, j $\in \{1, 2, ..., m\}$, i $\neq$ j. If all the blocks are disjoint, $B_i^\mu \cap B_j^\mu = \phi$ for any i, j $\in \{1, 2, ..., m\}$, i $\neq$ j cover μ is also called a partition on S^Z.

For two covers $\mu = \{B_1^\mu, ..., B_m^\mu\}$ and $\tau = \{B_1^\tau, ..., B_t^\tau\}$, the product $\mu \cdot \tau$ is defined as the set of all intersections of blocks of μ and τ:

$$\mu \cdot \tau = \{A \mid A = B_i^\mu \cap B_j^\tau \text{ for every } i \in \{1,...,m\}, j \in \{1,...,t\}\}$$

Note that $\mu \cdot \tau$ need not be a cover, since it may contain the void element. A case of interset arises when $\mu \cdot \tau$ is a partition, whose blocks contain exactly one element: in fact, every state $s \in S^Z$ can be assigned by specifying two blocks B_i^μ, B_j^τ, such that

$B_i^\mu \cap B_j^\tau = \{s\}$. Such a situation is denoted by: $\underline{\mu \cdot \tau = 0}$.

Working towards a decomposition of Z, a parallel decomposition δ^{PQ} of the input set X^Z is defined, as a one to one function from the pairs of elements of two sets X^P, X^Q to the elements of X^Z :

$$\delta^{PQ}: \ X^P \times X^Q \rightarrow X^Z.$$

A cover μ is consistent with set X^P of parallel decomposition δ^{PQ}, if, for any $\bar{s} \in S^Z$, any $\bar{x}^P \in X^P$, and all the elements $x^Q \in X^Q$, it results:

$$f^Z(\bar{s}, \delta^{PQ}(\bar{x}^P, x^Q)) \ \in B_j^\mu \ , \qquad\qquad B_j^\mu \in \mu.$$

A cover μ is closed with respect to set X^P of a parallel decomposition δ^{PQ}, if, for any $\bar{x}^P \in X^P$, any block $B_i^\mu \in \mu$, all the elements $x^Q \in X^Q$ and all the states $s \in B_i$, it results:

$$f^Z(s, \delta^{PQ}(\bar{x}^P, x^Q)) \ \in B_j^\mu \ , \qquad\qquad B_j^\mu \in \mu.$$

Examples of the above definitions can be found in the stationary machine Z, defined by:

$$X^Z = \{z_1, z_2, z_3, z_4\}$$
$$S^Z = \{s_1, s_2, s_3\}$$
$$s_1^Z = s_1$$

(8)

$s^Z \diagdown x^Z$	z_1	z_2	z_3	z_4
s_1	s_1	s_2	s_1	s_3
s_2	s_2	s_2	s_3	s_3
s_3	s_2	s_1	s_3	s_1

$$f^Z$$

A cover μ, and a partition τ, on S^Z, are respectively:

(9)
$$\mu = \{\{s_1, s_2\}, \{s_1, s_3\}\} \ ;$$
$$\tau = \{\{s_1\}, \{s_2, s_3\}\} \qquad .$$

The product $\mu \cdot \tau = \{\{s_1\}, \{s_2\}, \{s_3\}\}$; that is, $\mu \cdot \tau = 0$

Given two sets $X^P = \{x_1, x_2\}$, $X^Q = \{y_1, y_2\}$, a parallel decomposition of X^Z is:

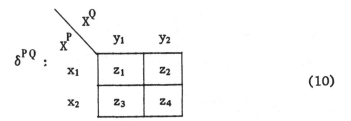

$$\delta^{PQ} : \qquad \begin{array}{c|c|c|} & y_1 & y_2 \\ \hline x_1 & z_1 & z_2 \\ \hline x_2 & z_3 & z_4 \\ \hline \end{array} \qquad (10)$$

By inspection of the table defining S^Z in (8), it can be verified that τ is consistent with X^Q, and μ is closed with respect to X^P.

We are now in the position of stating a decomposition theorem for a stationary machine Z.

Theorem — Let δ^{PQ} be a parallel decomposition of X^Z; $\mu = \{B_1,...,B_m\}$ be a cover on S^Z, closed with respect to X^P; $\tau = \{B_1,...,B_t\}$ be a cover on S^Z consistent with X^Q; $\mu \cdot \tau = 0$. Then, Z can be decomposed into a t.v.m. T with $X^T = X^Q$, and $\#(S^T) = t$, controlled by a monitor M with $X^M = X^P$, and $\#(S^M) = m$.

A constructive proof will be given for the theorem, by indicating how M and T can be formed. Since $\mu \cdot \tau = 0$, any state $s_h^Z \in S^Z$ is determined by a pair of blocks B_i^μ, B_j^τ, such that $s^Z \in B_i^\mu$, $s^Z \in B_j^\tau$.

The states s_i^M of M are put into correspondance with the blocks B_i^μ of μ, of same index i. Similarly, the states s_j^T of T are put into correspondence with the blocks B_j^τ of τ. Then, there is a correspondence between pairs s_i^M, s_j^T and states s_h^Z. In particular, the pair (or a pair) corresponding to the initial state of Z gives the initial states of M and T.

Since μ is closed with respect to X^P, a state transition for Z leads the machine from $s_h^Z \in B_i^\mu$ to $s_k^Z \in B_v^\mu$, where B_v^μ is determined by B_i^μ and $x_w^P \in X^P$ only. Then, the next state function for M is formed by letting: $f^M(s_i^M, x_w^P) = s_v^M$.

Since τ is consistent with X^Q, a state transition for Z leads the machine from s_h^Z to $s_k^Z \in B_v^\tau$ where B_v^τ is determined by s_h^Z and $x_w^Q \in X^Q$ only. s_h^Z is in turn determined by s_i^M and s_j^T. Then, the next state functions for T are formed by letting: $g(s_i^M) (s_j^T, x_w^Q) = s_v^T$.

The theorem can be applied to machine (8), with reference to covers (9) (τ, in particular, is a partition) and parallel decomposition (10). If states m_1, m_2 are orderly associated with the blocks of μ; states t_1, t_2 are orderly associated with the blocks of τ; the machines M and T result:

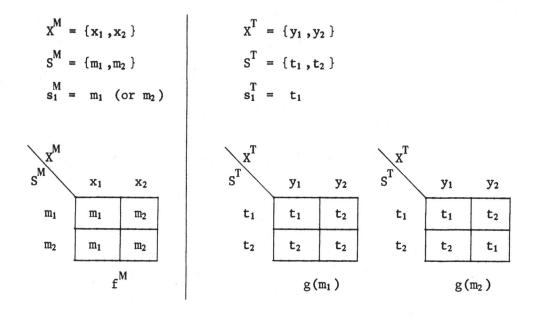

$$X^M = \{x_1, x_2\} \qquad\qquad X^T = \{y_1, y_2\}$$

$$S^M = \{m_1, m_2\} \qquad\qquad S^T = \{t_1, t_2\}$$

$$s_1^M = m_1 \ (or \ m_2) \qquad\qquad s_1^T = t_1$$

f^M $\qquad\qquad$ $g(m_1)$ $\qquad\qquad$ $g(m_2)$

3. Concluding remarks

The theorem presented in the previous section may be included in a research line, started by Hartmanis with the loop free decomposition of finite state machines. The original studies considered "autonomous" monitors with exactly one input x^M, $X^M = \{x^M\}$, such that any input sequence is a string of x^M, merely scanning the subsequent instants of time. The theory of such monitors, originally based on a partition algebra, has been subsequently extended to covers on S^Z [5]. Then, guided monitors in a broader sense have been introduced in [6], but the theory was still limited to binary machines with special partition properties. All such cases appear as particular instances of the decomposition theorem presented here.

A further (final) extension seems to be possible, if a set of covers on the input alphabet is considered, instead of a parallel decomposition of X^Z. It is worthwhile to note, however, that a parallel decomposition may spontaneously arise when the elements of X^Z are coded by the combination of values of a set of input variables.

REFERENCES

[1]   GINSBURG S., "Introduction to mathematical machine theory,, Addison-
      -Wesley, Reading, 1962.

[2]   GILL A., "Time varying sequential machines,, Journ. of the Franklin Inst.,
      267, 1963, pp. 519-539.

[3]   AGASANDJAN G.A., "Automata with a variable structure,, Dokl. Akad. Nauk
      SSSR, 174, 1967, pp. 529-530.

[4]   SALOMAA A., "On finite automata with a time variant structure,, Informa-
      tion and Control, 13, 1968, pp. 85-98.

[5]   GUHA R.K., YEH R.T., "Further results on periodic decomposition of
      sequential machines,, Proc. 3rd. Annual Princeton University Confer-
      ence on Information Processing Systems, Princeton University, 1969.

[6]   LODI E., LUCCIO F., "Automi finiti a struttura variabile,, Atti II Congresso
      Nazionale di Cibernetica, Casciana Terme, 1972.

AN OPTICAL SENSOR FOR LOCATING OBJECTS WITHOUT MARKS OR CONTACT

Eike MUEHLENFELD, Dr. Ing.
Institut für Informationsverarbeitung
in Technik und Biologie der
Fraunhofer-Gesellschaft e. V.,
Karlsruhe, West-Germany

(*)

Summary

Robots require optical sensors to recognize and locate objects under various conditions. The principles, operation and performance of a correlation sensor are described, which performs the cognitive computations by processing the light rays emerging from the object to be handled. Its simplicity may render automatic acquisition of all three position coordinates and one angle economical for many new tasks.

(*) All the figures quoted in the text are at the end of the lecture

1 Introduction

Automation through its progress in industry and administration is entering fields where the ability of man to perceive patterns and to recognize and locate objects is of utmost importance. Research activities in the field of automatic pattern recognition are aiming in that direction.

In production plants the primary interest is to replace human manual skill by mechanical manipulators and robots – in particular dangerous places for instance. These manipulators have to grasp a piece and to place it exactly into subsequent positions for machining, according to a given program. If such a manipulator is blind there is the problem of finding the object the manipulator is supposed to grasp. This problem can be solved by mechanical feeding devices which conduct the pieces into a precisely defined starting position, where the manipulator can grasp it blindly. Such a solution is feasible in some cases, but it is too expensive in other cases or not versatile enough to allow for frequent changes in the production process.

A picture processing system, an optical sensor with the ability to take over some of man's cognitive faculties, could enable a robot to distinguish between different classes of objects and to tell the position of the particular object to be handled. Corresponding electrical signals generated by the optical sensor could control the grasping, eliminating the necessity for precise prepositioning of the piece. In addition the optical sensor could control the final positioning and alignment at the particular machining stages. Such a picture processing system must remain within economical limits. A system requiring a large computer is beyond this discussion. We feel that our methods of optical picture processing – processing by optical means using the ordinary light emerging from objects – can serve as a basis for the development of suitable sensors. Practical results achieved with a sensor head, which had been tailored to somewhat different tasks, will demonstrate the feasibility of such a concept.

2. Specification of sensor requirements

Depending on the particular task and workshop environment, the requirements on the optical sensor will differ significantly. These requirements can be specified in nine categories with regard to the complexity of the perceptual task. The specifications concern the number of different types of pieces or objects, fed to the system in unpredictable sequence, and they concern the translational and rotational degrees of freedom of the objects. Position and/or orientation can be fixed by the feeding device. in which case a 1 appears in Table I, or it can be arbitrary. Because of considerably different perceptual problems we have to distinguish between angular freedom within the plane perpendicular to the axis of observation, and the freedom of tilt with respect to that plane.

The nine categories are specified, arranged in an order of increasing difficulty:

1. Presence-checking: The simplest sensory task occurs when single pieces of one kind are fed to an exact position in only one possible orientation. The sensor has to check the presence of a piece and to initiate a fixed manipulation program.

2. Sorting: Objects of different types in a fixed translational and angular position have to be recognized. The sensor has to initiate one of several manipulation programs according to the result of the classification.

3. Position fixing: The object is fed on a table or conveyer belt in arbitrary position but in a single angular orientation, defined by ajustable stops or magazines. The sensor has to locate the object to control the manipulator's grasping.

4. Sorting and position fixing: A combination of cat. 2 and 3.

5. Grasping from a table: Objects of a single type have a distinguished surface in flat contact with the table or conveyer belt. There are no mutual maskings. The sensor has to measure one angle in addition to position for manipulator control.

6. If the object has a finite number of surfaces which it may lay on, different aspects of the same object are to be treated as different types of objects (cat. 2), requiring the same or different manipulation.

7. Grasping out of a box: The sensor has to control the grasping of objects in arbitrary position and orientation, possibly masking one another.

8. The objects in the box are of different types.

9. Intelligent control: The most general and most difficult task, not required from industrial robots, is to process patterns of unlimited variety with poor a-priori-knowledge. Such situations are encountered in space or underwater applications or in catastrophe missions.

Sensors solving these tasks differ by orders of magnitude in the required technical and financial effort, convering the range from "blind" via "cognitive" to "intelligent". Their costs have to be compared to the costs of mechanical devices or organizational measures which might reduce the category of perceptual difficulty.

Tasks of category 1 have been solved with simple sensors like light-barriers, and need not bother us here. Tasks 7 through 9, dealing with 3-dimensional objects in arbitrary orientation, have found hopeful theoretical approaches and computer solutions, but are as yet rather far from applicability to the average production line. If not impossible, it will be cheaper to lower the category of difficulty into the region between no. 2 and 6, where economical solutions appear to be feasible.

3. Incoherent optical correlation

Tasks of category 2 to 6 consider the object as a 2-dimensional pattern, to which existing theories and techniques of automatic pattern recognition can be applied. Economical feasibility of the picture processing, as required, is achieved by performing the necessary computations by use of the many rays of light emerging diffusely from the object, represented by the black key in fig. 1. The object's light intensity is mathematically described by a function $f(x, y)$ depending on two local variables (x, y). There is a reference mask of the object contained in the sensor, a photographic negative in the simplest case, described by a transparency distribution $m(x, y)$. We want to know if and where object and reference mask coincide. The mathematical computation which yields this information is the cross-correlation between object $f(x,y)$ and reference $m(x, y)$:

$$(1) \qquad k\ (\xi,\eta) = \iint f(x - \xi,\ y - \eta)\ \ m(x,y)\ dxdy$$

For a few necessary modifications and additional computations the reader is referred to the literature [1]. The cross-correlation is exactly what we get by the simple setup of fig. 1. To understand this, we follow one of the diffuse rays of light, the one emerging from the rear end of the black key e.g., carrying f at a particular point (x, y) and proceeding parallel to the optical axis. Having penetrated the mask the ray's energy has been multiplied by the mask's transparency m at the same coordinates (x, y). All axis-parallel rays thus carry products $f(x, y)\, m(x, y)$ at respective coordinates (x, y) and are collected by the lens into its focal point, where the integral, one point of the correlation function k for displacement $\xi = \eta = 0$, is observable. In fig. 1 no light will reach the focal point, as axis-parallel beams either have zero energy, due to their origin from the black key, or cannot penetrate the mask. Other points of the focal plane do receive light by rays emerging from object points displaced by ξ and η . These rays are titled against the optical axis and penetrate the mask at undisplaced points (x, y). So in the focal plane we have the entire two-dimensional correlation function for all shifts simultaneously, which in our case shows a black dot on bright background. If the object is shifted to the border-lined position, the correlation dot is displaced accordingly.

The contrast of the correlation-dot is a measure of correspondence between object and mask. The correlation can be performed with several masks of different objects simultaneously. Such a multichannel-correlation is performed by the IITB's character-reader [2]. Comparison of dot contrast allows recognition and the control of sorting manipulations. The position of the correlation-dot gives two position coordinates in the object plane.

The principle of optical correlation has been well understood for a few decades. The problem encountered with ordinary incoherent light is the signal being carried by light-energy, which cannot become negative. Super-position of merely positive signals leads to an intensity distribution where the autocorrelation consists of a small peak riding on a high bias level and its noise. This problem must be coped with. Picture processing by coherent light would require transparent objects which can be transluminated by laser light, and is thus not applicable.

4. The optical sensor

The optical sensor is shown in fig. 2. In the orientation shown it looks towards the observer and correlates him with one reference mask behind the lens-system. In the housing to the right of the figure photoreceptors and their preamplifiers are visible. In the housing to the left we look via a mirror into a lens which forms an intermediate image f(x,y) which is used for correlation instead of the object itself.

The mirror is mounted on a swinging lever driven by alternating current. This introduces periodic shifts of the object-image and results in linear oscillations of the correlation-dot as shown in fig. 3, moving it across a photo-receptor. The extreme points of the dot's path are indicated by the arrowheads. Symmetrical excursions relative to the receptor (fig. 3a) cause the signal of fig. 4a with a fundamental wave indicated by the dotted line. It was twice the oscillating-frequency, as the correlation-dot will move to and across the receptor during one period. Such a symmetric excursion prevails for a prefered object position. Translation of the object in the direction of the path of the correlation-dot will cause turning points located unsymmetrically as in fig. 3b. The peaks in the signal approach one another by pairs producing the oscillating frequency itself as fundamental (fig. 4b). The signal energy is thus shifted from the first harmonic to the fundamental frequency due to object translation. This can be evaluated by phase-sensitive filtering, which at the same time allows a considerable bandwith-compression to reduce noise and drift of receptors, eliminating the problems of incoherent processing mentioned above.

The sum of spectral energy at fundamental and first harmonic frequency is proportional to the amplitude of the correlation peak and thus is a measure of resemblance between object and reference mask. To fix the second coordinate in the object plane the receptor is partitioned into two halfs as in fig. 3c. There is a difference signal if the correlation peak does not cover both receptors equally during cross-over. This signal is filtered accordingly, considering the sigh of the difference.

To acquire the third coordinate in the direction of the optical axis we utilize the intermediate imaging of the object into a plane which remains fixed within the limits of focal depth. Fig. 5b includes the oscillating mirror in front of the imaging lens, which produces the periodic movements of the correlation point mentioned above. It has been assumed, so far, that the image $f(x,y)$ and the reference mask $m(x,y)$ have the same scale, as in fig. 5b. If the image is larger than the reference mask, as in fig. 5a, rays of light, passing corresponding points in both, will converge ahead of the integrating lens, and will therefore produce the correlation-dot ahead of the focal plane. Receptors behind the focal plane will therefore receive a smaller share of the light energy than will receptors ahead of the focal plane. If the image is small compared to the reference, the correlation-dot will arise behind the focal plane, and the light energy will be distributed in a reciprocal manner, fig. 5c. As the first receptors should not shadow the rear ones, a beam-splitter behind the integrating lens generates two equal fields of radiation for each pair of receptors. The signal difference between rear and front receptors, filtered as described above, indicates the difference of the scales of image and mask. The scale of the image depends on the size of the object or on its distance from the imaging lens. This methods thus permits measurement of either size or distance of an object.

The angle between object and reference can be measured, too. We assume the mirror to be resting, for the time being. Rotational oscillations of the reference mask about the optical axis cause a signal according to fig. 4a, if there is no angular tilt of the object against the initial orientation of the oscillating reference mask. Angular tilt causes a signal according to fig. 4b and can be measured by the same signal processing, used for evaluation of object position. With different frequencies of mirror and reference mask both oscillations may occur simultaneously, if sum and difference frequencies are kept out of the signal bands.

5. Experimental Results

The essential features of performance of a position sensor can be specified by the precision of position fixing, response time and illumination required. Another important feature is the size of the field of target acquisition. These performance data are interdependent and can be exchanged mutually within wide limits by proper choice of parameters, like mirror amplitude, filter bandwidth and the distance between image and reference mask.

Most important for sensor performance is the structure of the object — its topological information content — and the contrast within the object and with respect to its background. The position of richly structured objects, like printed circuit boards, in the plane perpendicular to the line of sight has been measured with a reproductibility of $\pm 1 / \mu m$ in a distance of 20 cm, corresponding to an angular resolution of ± 1 arc sec. Response time is limited by the vibration frequency of the mirror to 5 msec. The field of acquisition has a diameter of 2 cm and can be extended to an angular diameter of $\pm 10°$. Theory yields that coarsely structured objects require proper filtering in the spatial frequency domain, which in many cases can be described as a simple contour-reduction of the mask. A quantitative description of structural fineness and its influence on sensor performance are not yet available.

6. Flexibility of sensor performance

An optical sensor of the type described yields electrical signals for the resemblance between object and reference mask and for three position coordinates. An ancillary device measures the angular orientation in the object plane, too. These analogue voltages can be fed to a manipulator for automatic control of grasping and positioning. The range of the object can be choosen deliberately by distance adjustement of the imaging lens. There is no special light source required with ordinary daylight. The sensor operates without contact and without special markings on the object, as long as its contours are clearly distinguished from its background, the conveyer belt e.g., or the object itself is structured sufficiently.

As the entire pattern of the object is evaluated for correlation, even comparatively large spreads between objects can be tolerated. A transport device for the mask-film can match the sensor to changing tasks within a second.

The optical correlator does not meet all possible demands, and there will be many cases where it is inadequate, because the object does not lend itself to correlation techniques or because the category of perceptual difficulty, as specified in table I, is too high. But its simplicity and versatility may render robots economical for many new tasks.

REFERENCES

[1] MÜHLENFELD, E., Parallele Verarbeitung zur Erkennung beliebiger Bild-
     muster, Nachrichtentechnische Zeitschrift, NTZ, 1970, pp. 597-603.

[2] HOSAGASI, S., Optischer Korrelator zum Zeicheneinlesen in Rechner,
     Mitteilungen 1972/1973, Institut für Informationsverarbeitung in
     Technik und Biologie, pp. 18-23.

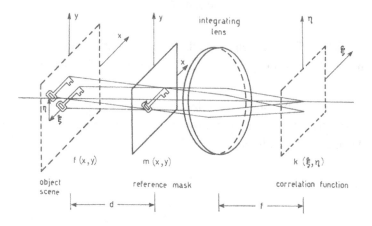

Fig. 1. Incoherent optical correlator for twodimensional patterns

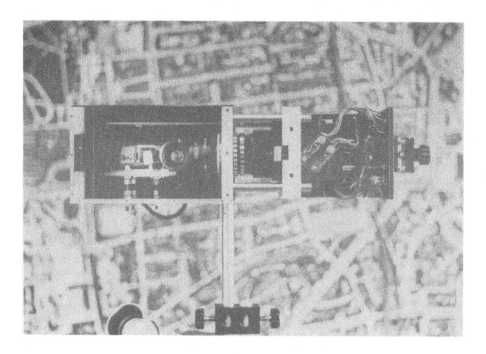

Fig. 2. The optical sensor for recognition and position fixing consists of an oscillating mirror, 3 lenses, the reference mask and 4 receptors behind a beam-
-splitter.

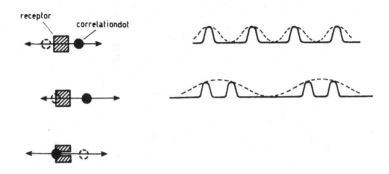

Fig. 3. Path of the correlation-dot with mirror oscillations

Fig. 4. Receptor signals as function of time with different object positions

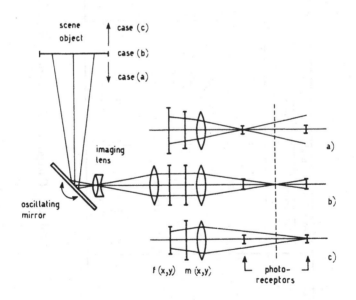

Fig. 5. Formation of correlation-dot with different ranges of the object.

Category	Work pieces, number of		possible angular orientations		Robot function
Specif.	different types	possible transl. positions	in plane	tilted	
1	1	1	1	1	Presence checking
2	finite	1	1	1	Sorting
3	1	arbitrary	1	1	Position fixing
4	finite	arbitrary	1	1	Sorting and position fixing
5	1	arbitrary	arbitrary	1	⎫ Clutching from a table
6	finite	arbitrary	arbitrary	finite	⎭
7	1	arbitrary	arbitrary	arbitrary	⎫ Clutching out of a box
8	finite	arbitrary	arbitrary	arbitrary	⎭
9	arbitrary	arbitrary	arbitrary	arbitrary	Intelligent Robot

Table 1: Categories of perceptual difficulty for optical sensors.

SENSOR DEVICES FOR INDUSTRIAL MANIPULATORS AND TOOLS

J.R. PARKS
National Physical Laboratory
Teddington, England

(*)

Summary

With the gradually increasing application of industrial manipulators and robots the need is emerging for sensory devices capable of providing these devices with a considerable degree of contact with their environments. Existing use of simple contacts and photo-electric detectors are limited to detection of limit stop conditions or obstruction. The several advanced robot projects are aimed at producing almost human sensory capabilities and are too slow and expensive for industrial application.

Between these two extremes there is a range of possibilities for development of sensory devices of economic significance through the use of artificial environment conditions (of ambient lighting etc.). Some tasks of pica-part inspection and location can be accomplished using available technology, some indication of which is given in this paper. The paper discusses briefly some tactual and visual image approaches to sensory systems. An apprach to machine monitoring through machine reaction is discussed.

Stress is laid on the need for evolutionary, rather than revolutionary, development of machine sensors in order to obtain users confidence in production application.

(*) All figures quoted in the text are at the end of the lecture.

1. Introduction

With the exception of mechanical or opto-electronic limit switches commercially available industrial robots or universal transfer devices (UTD) are in the literal sense senseless. This has not, as yet, proved a fundamental block to the application of UTD's, as they are far from being fully applied in simple, well controlled situations where dead reckoning is practicable.

However, as UTD's become more familiar and cheaper then the range of applications will increase and the need for sensory devices will becomeapparent.

It is the purpose of this paper to indicate some of the potential developments of practical and economic sensory devices. The attitude of the paper derives from a background in the development of practical pattern recognition devices.

UTD's are generally used for picking objects from one area and placing them in another. Variations of object indentity or quality, position and orientation are all potentially critical in various combinations and four distinct sensory requirements emerge:

1. To discriminate different objects, either to sort or, more simply to discard all objects not of a predefined type.
2. To measure the position and orientation of a known object. Some knowledge of the general shape of an object is necessary to make meaningful position and orientation measurements.
3. To examine a known object and assess its quality, e.g. missing or misplaced piercings, surface finish, clearance of burrs and flashings etc. precision measurement.
4. To guide a manipulator in placing objects relative to previously placed objects, i.e. assembly or packing.

2. Approaches

The experimental robotic groups in Japan, United Kingdom and USA are well known [1,2,3], but from the commercial and industrial point of view these are scientific toys. However, it is from these sources that many of the basic ideas must emerge which will eventually lead more commercial engineers to devise sensory systems for use in production situations.

The abstract nature of some robot projects centres around the project groups motivation to evolve devices capable of becoming man-substitutes for working in natural environments. The production factory environment does not have to be natural or conventional and there is freedom, within economic bounds, to impose whatever environmental conditions are most useful. For example, to inspect instrument movement plates which have been coined, visual inspection using frontal lighting of the parts distributed on a wooden bench would be unnecessarily difficult. The textures of the wooden bench and the surface marks of the parts contribute to a complicated image when the information in which we are interested is conveyed by the outline of the plate and its piercings.

A more practicable approach would be to view the plates placed singly on a self-luminous surface and examine the silhouette obtained. This and other methods for simplifying the environment are discussed in later sections.

It is conventional to think in terms of sensory devices attached directly to the manipulator, this is not always necessary nor even desirable. For instance one of the major weaknesses of automatic assembly devices centres around their inability to detect and/or cope with defective piece-parts. This trouble can be overcome if the acceptable quality level AQL, is improved from the prevailing 1/2 to 2 per cent for low-cost high-volume piece-parts and fastenings to the level of 0.1 per cent or better, essential for substantial cost effectiveness of automatic assembly systems. This improvement can only be achieved economically using automatic means. This degree of inspection has to be achieved while introducing a mark up of only 5 to 10 per cent in component cost.

Similarly problems of piece-part alignment (for assembly) could be reduced in many instances through the use of magazined components, using sensory devices to direct simple manipulators to load magazines automatically at a point independant of the eventual assembly operation.

The major advantage which accrues from the separation of sensory devices and the manipulators which they serve is that both devices are free to operate at their maximum rates. Individual rates may well be quite different or individually variable and the cost effectiveness of a system in which both devices are directly linked would be compromised, through mutual interference.

So far we have not defined what sensory capabilities we are considering. Because they are already very familiar, direct measuring devices, such as rotational or linear encoders and simple go/nogo limit or obstruction detectors will not be discussed. The range of sensory devices then left to us is considerable and can be classified according to the form of the physical property being sensed.

These are in general limited to:

(a) two dimensional images of a two or three dimensional scene.

(b) one dimensional analogue variables with time, derived from a continuous process.

For class a) transducing devices such as TV cameras, possibly working in selected parts of the EM or sonic spectrum are available and familiar as image devices. Also included are tactile sensors which can be used to contact objects and thence transduce a 'touch image'. Into class b) fall a large range of devices: thermometers, microphones, accelerometers, strain gauges, pressure gauges, flow gauges etc. Many of these devices have a rate of response short compared with process variables and can be used for continuous monitoring.

Both classes of transducer are potentially valuable in increasing the sensory contact between a machine and its environment. The central problems concern how to use systems with sensory capability in a commercially effective way.

3. Visual sensors

There are several examples of work on the interpretation of natural, if somewhat contrived, visual scenes [4,5] but rather less on the manipulation of deliberately contrived scenes with minimal resources [6,7]. Minimal resources implies a small computer with a limited amount of special purpose equipment attached. The current natural-scene analysis programs take many seconds to process a simple scene using very large computers. Such facilities obviously fall outside the cost constraints indicated earlier by several orders of magnitude.

The situation improves rapidly upon reduction of initial aspirations. Consideration of a) means for reducing the information content of scenes to a level compatible with available technology, and b) what visual tasks can be approached from a strictly limited equipment cost point of view produce a changed attitude to the problems for which solution is attempted.

As mentioned earlier, back lighting arrangements or coloured support surfaces give a simple black and white scene in comparison to the complex grey scale scene obtained by using frontal lighting. Every picture element in the scene theoretically has only two possible values and the amount of information adequately representing the scene is drastically reduced (perhaps by a factor of 8 to 10), the management of this information is correspondingly eased.

Having reduced the scene to binary from only the points representing the black-white transition boundary are required to define the shape. Computers as currently engineered are not suitable for handling two dimensional imagery, though specific devices are under study [8]. However, it is possible, at moderate expense to devise hardware systems which will take the output from a suitable scanning device, such as a closed circuit television camera and produce a description of an object in terms of a list (carefully ordered) of the edge points of its silhouette.

Though the TV camera is scanning a silhouette it is, in practice, difficult to maintain the level and the uniformity of the background luminosity constant. The output of the camera, therefore, has to be treated as a continuously variable (ambient) waveform superimposed upon a binary waveform representing the image. In the presence of 'shading' in the optical source scene and non-uniformities introduced by both the lens system and the camera tube we wish to recover a dimensionally accurate representation of the object. A simple fixed threshold clip is far from adequate, and alternative methods are required.

Such a system, indicated in Fig. 1. is based on a method of line extraction described by Shirai [9].

The bounds of an object are optimally defined by the point of inflexion in the brightness distribution around and close to the edge of an object. The system shown detects this inflexion without loss of resolution of the image and produces a code representing, in readily computable form, the image discontinuities. This 'chain code' can be reduced by relatively straight forward processes to give a description in terms of line segments specifying the object boundaries, including any piercings. Existential and dimensional checks can then be performed while manipulating a minimum amount of information. The storage organisation for most of this processing is in the form of ring structures with some list structuring which will be controlled by a micro-computer.

The minimal usable picture fragment for purposes of determining slope and slope direction is a 2 by 2 picture element window. Given this it is possible to estimate the brightness slope or gradient and also the direction of the slope (i.e. the direction of greatest increasing slope).

These quantities are significantly nonzero only at or near brightness discontinuities. At all other points, either in background field or on the object. the brightness gradient is negligible. The storage required to retain the description of the object is relatively small and the data rate is also reduced from the original TV picture continuous element rate of some 7 MHz to asynchronous brusts as a boundary is crossed. The true boundary corresponding to the inflexion in the brightness is defined as the greatest slope encountered in the mean slope direction. This method retains the resolution of the original image without significant tendency to round off sharp corners.

The use of a micro-computer (computer-on-a-chip) allows ready implementation of more complex conditional calculation upon the local slope and slope-direction than is practicable with a hardwired system of economic proportion. The relative inability of conventional computing systems to handle two dimensional imagery economically is one of the primary obstructions to the development of economic artificial visual systems, the use of micro-computers controlling specially contrived storage arrangements will offer solutions to these problems.

Once reduced to simple line structure description comparison with acceptable standard forms is required. Though still in itself quite a complex task, it is appropriate for solution by conventional mini-computers in which manipulation of symbolic descriptors is readily achieved.

Schemes such as that outlined above will be capable of processing several images per second of a complexity typical of small punched components.

Once developed and understood an extended philosophy will be capable of dealing with selected, front lit, situations. For example, inspection of integrated circuits or printed circuit boards, (still essentially planar objects) possibly partially assembled. Some work in this area has recently been reported from Hitachi [10].

The above account indicates one attitude to the visual sensor. There are, however, altenative techniques which also exploit the opportunity presented to modify the environment in a production situation. Principle among these is the use of 'structured illumination' described by Pennington and Will [7]. In their approach a regular mesh pattern or grid is imaged onto a scene containing three dimensional objects. If this scene is obliquely lit with respect to the camera, then the various plane surfaces in the scene have a grid structure imposed on them. This is characteristic of their attitude and range relative to the illuminating projector Fig. 2.

This approach or modification of it appear to be particularly useful for tasks such as package handling.

Since the resultant illumination structure is periodic and Fourier techniques are appropriate for the detection of plane surfaces. The areas illuminated are extensive relative to the brightness discontinuities existing at boundaries in uniformly lit scenes (in typical lighting condition these discontinuities can be vanishingly small). Also non-periodic structure on the surfaces, e.g. labels, advertising material etc., is partially suppressed.

Another method of exploring a scene which has been described is the projected line [11]. In this method a single extended straight line is projected upon a scene, otherwise in darkness, which is viewed by a camera placed at a suitable angle to the direction of projection Fig. 3. As seen by the camera the projected line is bent around the objects in a way characteristic of the shape of the objects (and a function of the relative position of the line projector and viewing camera and the position and attitude of the object).

A single slit position only illuminates a small fraction of the scene; a complete scene is explored by scanning the slit across the scene in a direction normal to the slit. This method is apparently rather slow requiring incremental stepping of the slit over the field of view. This apparent disadvantage may be set against the relatively small amount of information which is handled at any one instant. This system has rather less immunity to surface defacing than the system of Pennington and Will using a grid illumination pattern.

The main interest in both the 'grid' and 'strip' methods is their relative independence of edge detail and surface markings.

In the grid illumination system only those areas having an extent of several periods of the incident structure are reliably detectable, similarly curved surfaces, though producing some pretty effects, pose non-trivial aperiodic analysis problems ! Thus the resolvable detail possible with mesh illumination systems is limited. This restriction could be alleviated to some extent by mounting the visual system (projector or camera, or both) on the moving hand of a UDT. As objects are approached then the period of the projected structure would decrease thus increasing the available resolution.

Strip illumination methods are less dependant upon surface extent but are inherently rather slow if high resolution is required.

While the universal 'seeing' system is still very far off it is to be expected that limited capabilities implicit in the simple visual sensor discussed will not only be useful in themselves but that they will lead, generically, to the development of more advanced capabilities. Such has been the experience in a number of fields of endeavour in the past.

Higher order visual capabilities are, for the moment at least, the domain of the AI projects described earlier. The problems of watching a manipulator and controlling it as it attempts to assemble objects is daunting. (There is an account of a brick stacking robot which 'looks' at the task to be done, makes an attempt, without visual sensing (i.e. with its eyeshut) and then inspects its achievment afterwards, having removed its 'hand' from the field of view [12]).

Continuous visual monitoring of the movement of a complicated 'hand' has, so far, not been demonstrated and the prospects for a continuously visually monitored assembly robot are not good for the forseeable future.

4. Tactile images

The development of tactual sensors for use in packing and assembly situations is receiving attention. Work of Larcombe [13] and Goto [14] has shown that sensory systems offering a limited two dimensional image of an object can be used to determine the exact relative positions of a known object and the 'hand' of the manipulator. This will be more accurate than a hand directed by a visual sensor alone.

Development of visual systems for approximate location of objects with fine location given by a tactile sensory system in the hand of the manipulator appear possible. Though the realization of such a concept is some way off.

In the short term tactual sensory systems could well find application in improved pick and place assembly systems in which piece parts roughly located and oriented by simple magazines, or more conventional automatic feeders, are handled in a precisely determined attitude through the use of tactual information. A dramatic increase in the feed rate of bowl feeders etc., can be expected when the associated placement device can cope automatically with a small but significant proportion of unacceptable presentations. The unacceptable condition could be due to mis-orientation or foreign bodies. Existing methods using various mechanical hazards and photoelectric detectors are all dependant on the rejection of specific non-allowed conditions; this is not equivalent to direct detection of allowable conditions since what is not allowed is inevitably a very large set of conditions which cannot be fully specified. Sorting by exclusion is therefore more prone to error than a direct identification of the correctness of a situation.

Both tactual and visual sensory devices make direct correctness checks possible, with consequent reduction in the problems of excluding unacceptable objects or presentations.

5. Non-image sensory systems

In seeking to improve the productivity of a modern manufacturing facility through the improvement of component quality we can usefully consider approaches other than post manufacture inspection of piece parts. It is obvious that substandard parts discovered some time after manufacture can only be scrapped, or returned for rectification. If defective pieces can be discovered at the time of production then production can be halted while the producing machine is examined to determine the cause of the poor quality. In this way defective production is stopped as soon as it begins with consequent reduction in level of scrap (which has been estimated to cost British industry more than £500 million per annum) and also increase machine utilization; it should also be possible to detect the on-set of machine malfunction before tooling etc., becomes irrecoverably damaged.

Point of production inspection or process certification can be achieved either by (visually) inspecting parts, while in the machine tool or alternatively we can monitor the production process rather than the product itself.

Inspection of piece parts while still held in the manufacturing tool is similar in principle to post manufacture (off line) inspection methods but has the advantage that the position and attitude of parts is known. However, the production and inspection processes have now to be related in their rates; and disparities will result in inefficiencies unless sophisticated systems for time multiplexing scanner heads etc are introduced. Trouble can also arise through swarf and lubricant adhering to the part.

Process certification is indirect and is not necessarily as effective as direct inspection. Indeed the technique would be directed towards maximizing machine tool utilization and reducing the probability of damaging and scrap producing situations from arising and remaining undetected.

We are not here concerned with adaptive control of machine tools though this could be considered to be an advanced derivative of process cerification. The objective here is to suggest low cost methods of process certification in order that alarms may be sounded to indicate the need for human operator intervention. This approach avoids the very difficult instrumentation problems in adaptive control needed to isolate the various process variables.

There is a noticeable tendency in recently developed manufacturing processes for the use of short duration high work-rate methods, for example, cold forging, extrusion of spur gear and similar complex components, high force presses. These tools form components from preformed metal blanks or billets in well defined, fixed number of stages. The high work rate of these processes produces a short lived but intense reaction in the machine tool itself. As the processes are well controlled this reaction is, under normal conditions, reproduced on each machine cycle. Thus we have the idea of a 'work function' for high rate processes. This function is arbitrary but constant for any correctly running machine. Significant variation from the normal shape of the work function can be detected and used as a warning of unacceptable process changes.

Fig. 4. indicates the form of an idealised work function and the effects of some variables.

Two points should be noted here a) since the work function is arbitrary it is not possible to identify the process variable in error, which could be material variation as well as process variation and b) if the work function changes then it is certain that the product will be incorrect, but some variation in the process may not be detected unless the reaction measured is carefully chosen even though the function shape is arbitrary.

For metal forming high rate processes such as those indicated earlier the work function is most easily obtained by strain gauges attached on or near to the tooling area of the machine. Accelerometers or microphones could be used but would be unduly influenced by vibration communicated from close cycle machines.

It is probable that similar functions can also be derived from many other short duration processes such as current profile in spot welders, metal flow profile in die-casting machines etc.

There are several possible methods available for comparing waveforms described in the literature. Straightforward cross correlation using the method of matched filtering [15] is conceptually simple but can be computationally expensive. If there is significant variation in machine speed (which only varies the period of the function without changing its basic shape) suitable normalising methods are necessary, again probably expensive and inconvenient. In the presence of speed variation alternative techniques are appropriate, for example, the method of zero-crossing is well known as the basis for comparing two complex wave-forms [15]. In this method the relative proportion of the waveform (or a derivative of it) between points at which it is zero valued are used as a basis for comparison. The exact method will vary with the application but the generality of the approach is only marginally affected.

In a typical system a processor can be used to monitor a shop of say a hundred cold forge heading machines operating at rates upto eight pieces a second. The on-machine equipment is limited to a single strain gauge with simple amplifier costing less than £20. Each machine output is connected to a central data scanner system under the control of a small computer. This also examines incoming waveforms and causes an alarm if any machine is malfunctioning. It is not intended that every single work cycle is a complex such as this would be examined but that each machine would be assessed at regular intervals so that any trouble will be quickly detected. It should be possible to detect a gradual drift in the work function so that advanced warning of imminent machine failure can be given.

The overall cost of such a system, produced commercially would be less than £10K. The basic system would allow supervision of say 100 machines as indicated above at a cost of the order of £ 100 per machine. Alternatively a group of 5 to 10 large presses operating at rates of a few cycles per minute could be monitored on each cycle at a cost of about a thousand pounds each. An errant and potentially dangerous or self-destroying process could then be automatically stopped after a single unsatisfactory work cycle.

6. Application

While many things are technically feasible it is appropriate to discuss shortly how visual inspection or process monitoring systems would operate in practice.

A central problem in application concerns the means by which the various systems could be instructed. Clearly any dependence upon detailed notational specification of piece parts or machine reaction would greatly reduce the practical applicability of the various systems and more amenable methods must be used.

In the case of process certification we can assume that, at the beginning of each production run, the machine will be correctly set up by a skilled tool setter. When he is satisfied that the machine is truly set the monitoring system is signalled and will then observe each of a statistically singificant number of successive machine cycles during which it will automatically derive the statistics of the process signal. Thereafter comparison of selected process signals with the acquired reference data would be totally automatic.

It is apparent that the success of such a scheme is no less dependant upon the skill of the tool setter than at present. We gain from the reduction in the dependence on itinerant machine minders to regularly and conscientiously check machine operation.

In the case of visual (and probably tactual) systems, at the simplest levels, a similar approach can be used. Since it is an image handling system then the most obvious way to describe an acceptable part is to show it one. Subsequent parts can then be compared with the certified reference and passed or rejected by the system according to the degree of match obtained. In practice we would not wish to accurately locate or orientate piece parts for inspection — the system outlined earlier is in fact capable of coping with objects in any attitude, provided that they are completely in the field of view.

At a higher level of inspection a range of tolerances acceptable at different points on an object can, in some instances, be indicated to a system by scale drawn profiles of the parts conforming to maximum and minimum limits. The system is then required to determine whether or not piece parts fall within the bounds.

Tasks such as the direction of a manipulator to pick up a part across particular 'flats' will initially, at least, require operator instruction via a simple visual display screen using a moveable cursor to indicate his wishes. Automatic flat seeking methods have not yet been demonstrated, but are known to be under investigation.

It is believed possible that tasks such as parts assembly, visually monitored, can be specified by an extension of the 'training by rehearsal' methods already familiar in the 'training' of many programmable UDT's.

Clearly the situation will be more complicated and less defined than in current practice and it is expected that early visual and tactual sensory systems will be employed to detect and reject as unsatisfactory situations in which gross errors are found (rather than try to correct them). Components would be returned for reuse thus avoiding potentially damaging or otherwise unsatisfactory situations at the cost of a lost work cycle.

It will be readily appreciated that, given sensory systems which can be 'trained', then they can just as easily be 'retrained' when a particular job is finished.

Such sensory systems are, therefore, part of the Programmable Automation attitude to production engineering and are suited to and complementary to the programmable robots now available in increasing number [16].

7. Conclusion

While not describing any specific sensory systems in detail several approaches to the development of useful sensory systems for use in industrial application have been indicated which give hope for the use of relatively advanced sensory devices in the short term. These approaches have been in part, at least, seeded by the several advanced sensory robot experiments in progress at several laboratories in USA, Japan, Great Britain and supported by a familiarity with some of the commercial constraints to be observed in developing practical pattern recognition systems.

By the use of carefully contrived situations, particularly of methods of illumination the possibility of improving the quality level of piece parts for subsequent automatics assembly and handling using is apparent. The use of contrived situations in order to simplify a task is not new in the field of production engineering and the use of special lighting effects etc., as required is limited only by the cost-effectiveness of the complete operation.

The eventual development of visual sensory capacity of some generality appears likely and desirable, it will probably be most effectively guided by an evolutionary development from simple devices. In this way successive generations of sensory devices will have respectable antecedents and be introduced into an informer market. It should not be assumed that seeing mechanisms of extensive generality will be available within the forseable future though several evolutionary steps from simple beginnings can be expected.

Tactile sensitivity is a valued human facility and we may expect to see a similar cooperative interaction between visual and tactual sensory systems used to serve precise positioning manipulators. So far there is no widely available tactile image scanners complementary to the range of optical image devices.

It is conventional to consider sensory systems as being directly linked to and interacting with manipulators. This is not necessarily the most effective use of manipulator or sensors due to the disparity in their operating rates. This can probably be overcome to some extent by time sharing the sensory system between several manipulators, though each will need its own scanner.

Separation of manipulator and sensory devices through use of suitable intermediary carriers, magazine systems etc. allows component inspection near to or even as part of the manufacturing process. Direct inspection of piece parts may not be essential and there are a number of processes which can be depended upon to produce good produce if their correct operation can be regularly certified. These processes are, initially at least, limited to those of short duration eg extrusion, forging, spot welding etc.

There are a large number of potential applications for the techniques briefly sketched above particularly in the large-volume low unit cost production areas such as fasteners, instrument manufacture and a wide range of automotive components. The techniques are not special purpose in themselves and are of considerable generality and offer a change of attitude towards many problems of inspection, machine monitoring and automatic assembly currently performed either manually or by expensive special purpose devices.

REFERENCES

[1]   Ejiri M., 'A Prototype Intelligent Robot that Assembles Objects from Plan Drawings.' IEEE Trans. on Computers. VOL C-21 No 2. Feb. 1972.

[2]   Popplestone R.J. 'Freddy in Toyland.' Machine Intelligence 4. Edinburgh University Press. 1969.

[3]   Feldman J., G.M. Feldman, G. Falk, G. Grape, J. Pearlman, I. Sobel and J.M. Tenebaum. 'The STANFORD Hand-eye Project' Proceedings of the First International Joint Conference on Artificial Intelligence. May 1969. pp 521-526.

[4]   Clowes M.B. 'On Seeing Things'. Machine Intelligence 2. Edinburgh University Press. 1965.

[5]   Winston P.H. 'The MIT Robot', Machine Intelligence 7, Edinburgh University Press. 1972.

[6]   Hegginbotham W.B., D.W. Gatehouse, A. Pugh, P.W. Kitchin and C.J. Page. 'The Nottingham SIRCH Assembly Robot.' Proceedings of 1st Conference on Industrial Robot Technology. International Fluidic Services Ltd., Bedford. 1973.

[7]   Will P. M. and K.S. Pennington. 'Grid Coding: a Preprocessing Technique for Robot and Machine Vision.' Proceedings of 2nd International Joint Conference on Artificial Intelligence. British Computer Society. 1971. pp. 66-70.

[8]   Levialdi S. 'Parallel Pattern Processing.' IEEE Trans. on Systems, Man and Cybernetics. VOL SMC-1 No 3 1971. pp. 292-296.

[9] Shirai Y. and S. Tsuji, 'Extraction of Line Drawing of 3-Dimensional Objects by Sequential Illumination from Several Directions.' Pattern Recognition. VOL 4 pp. 343-351 1972.

[10] 'Hitachi Robot Reads Patterns' Electronics. April 26 1973.

[11] Shirai Y. and M. Suva. 'Recognition of Polyhedra with a Range-finder'. Proceedings of the Second International Joint Conference on Artificial Intelligence. British Computer Society. 1971. pp. 80-83.

[12] Pringle K.K., J.A. Singer and W.W. Wichman, 'Computer Control of a Mechanical Arm through Visual Input' Proceedings IFIPS 1968. North Holland, 1968. pp. 1563-1569.

[13] Larcombe M. H.E. 'Tactile Perception for Robot Devices.' 1st Confernece on Industrial Robot Technology. International Fluidics Ltd., Bedford. 1973.

[14] Tatsuo Goto, Kiyoo Takeyasu, Tadao Inuyama and Raiji Shimomura 'Compact Packaging by Robot with Tactile Sensors.' Proceedings of 2nd International Symposium on Industrial Robots. IITRI Chicago. 1972.

[15] Ullman J.R. 'Pattern Recognition Techniques.' Butterworths 1973.

[16] Anderson R.H. 'Programmable Automation. 'Datamation Dec. 1972.

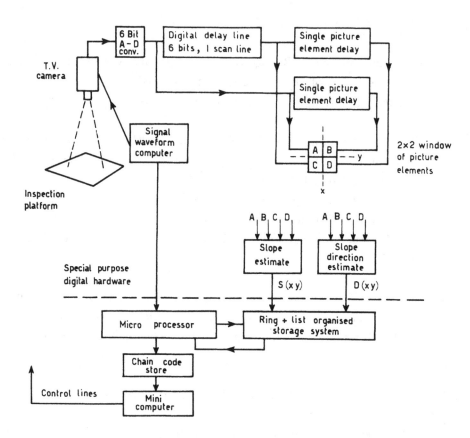

Fig. 1 Basic system organization for rapid scene analysis.

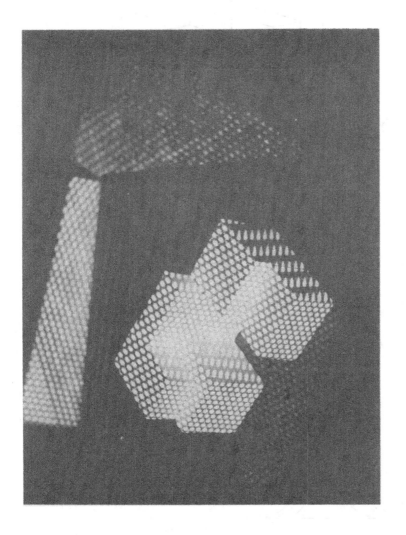

Fig. 2 A scene-of-blocks illuminated by a hexagonal "grid" structure. Note transformation of pattern on faces of different orientation.

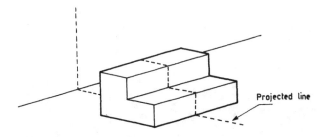

Fig. 3 Analysis of projected line-line projected overhead, viewed obliquely.

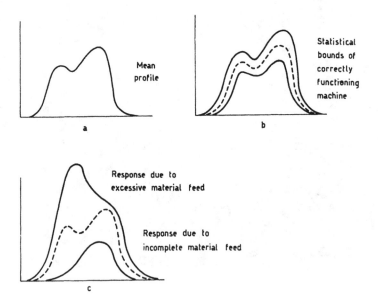

Fig. 4 Idealized strain curves typical of cold forging.

ON DEFINITION OF ARTIFICIAL INTELLIGENCE

RADCHENKO A.N.

master of science on cybernetics
Politechnical Institute named M.I. Kalinin
Leningrad, USSR

YUREVICH E.I.

Professor on cybernetics
Politechnical Instiute named M.I. Kalinin
Leningrad, USSR

(*)

Summary

The paper shows the efficiency of interpreting the behaviour of humans, animals, and robots as manifesting the relationship between two texts - afferent and efferent. The transformation of one text into the other is treated as quasi-linguistic translation. Intelligence of behaviour is defined in terms of exactness and brevity of translation.

A model is presented, as developed at the Polytechnical Institute (Leningrad), which affords the recording of the efferent text as a function of the informational structure of the afferent text. The read-out is executed by reproduction of the afferent text; if the latter consists of insignificantly modified fragments of the earlier text, fragments of the efferent text are reproduced at the output in a correspondingly modified sequence. The intelligence of the behaviour of this system is analyzed, and it is found to be higher than in the case of "word-for-word translation" since the model has capacity of detection and extrapolation, and also since there is no preliminary segmentation of texts into "words".

The linguistic approach suggests that it is not justified to assimilate the brain to the computer.

(*) All figures quoted in the text are at the end of the lecture.

Introduction

Research on artificial intelligence, conducted on a wide scale in many countries, has a characteristic feature distinguishing it from all other fields of science. It is the fact that the designers of "intelligence" have no definition of the object they want to build. This fact has repeatedly given occasion to justified criticism - indeed, how can one construct something one does not know ?

It is noteworthy that all results thus far achieved in - let us call it so - the field of "artificial intelligence" have been obtained by traditional scientific methods, implying that certain goals are conceived and defined in advance. All these achievements borrowed from well elaborated techniques of attaining a goal set beforehand. This pertains to studies in the field of second-generation industrial robots, pattern recognition, predictive and extrapolating systems; other instances are the concepts of self-programming systems, the existing theorem proving algorithms, etc.

What is the reason for referring these rather independent lines of research to the common field of "artificial intelligence" ? Is not this term just another word for cybernetics, which was defined by Norber Wiener as the science of universal regularities of control and communication processes in machines, living organisms and associations of these ?

The terms do not seem to be synonyms, although "intelligence" doubtlessly belongs to the terminology of cybernetics as it can be "artificial" or "natural" thus characterizing a certain property of a technical or living system.

The time is ripe for giving a definition of the term "intelligence" as used in a number of natural sciences (psychology, physiology). This defintion becomes absolutely indispensable, if we want to be wise in building wise machines. Lack of clarity may be pernicious to the development of "intelligent" robots.

In view of this, an attempt is made below to define the concept "intelligence" and to assess the immediate outlook for the development of intelligent automata.

1.Intuitive approach

An automatic device working according to a rigid program, however sophisticated, poses no question about intelligence. However, often the program goals and subgoals can be achieved in a multitude of alternative ways, more or less efficient. Therefore, we shall take into account this property in comparing intelligence levels. Still, the notion of "intelligence" is not indispensable for characterizing the "laconicism" of the machine's operation.

Ashby [1] reduces the notion of intelligence to the faculty for independent choice, stipulating that the latter is to be made on the basis of processing sufficient information for judging about the correctness of that choice. How can one define the criterion of correct choice ? The problem is reducible to the choice of the suitable criterion, that is to the same problem of choice but with additional information.

We should like to add that the term "intelligence" is reasonably applied only to automata operating in an unpredictibly variable environment. This term is appropriate in apprising the behaviour of the device in a changing medium. The assessment should be made dependent on the extent to which the behaviour is adequate to the current state of the medium - external and internal. The criterion of adequate behaviour in a changing environment is equally applicable to humans, animals, and robots.

Psychologists employ numerous tests to measure the intelligence quotient of humans and animals. The tests are a convenient method for comparative estimation of professional intelligence, and they are widely used. On the other hand, this method has been repeatedly and justly criticized: a person having a high I.Q. by one test is often of a low intelligence according to another test. A testing situation is imaginable where a gun dog gets a higher I.Q. score than its hunting master. Nevertheless, the method of tests can be of use for appraising the intelligence of automatic devices. It must be stressed, however, that exact evaluations can be obtained only by diverse testing of the automata in a real environment. In this way the information capacity of the device is established, and the "occult knowledge" eliminated from the estimate.

Another question arises: To what extent the environment can be variable ? In some cases the variations are just minor disturbances hampering the standard behaviour. In other cases the fulfilment of the standard response requires adaptation at the level of pattern recognition. In still other cases, the machine must, moreover, drastically modify its behaviour to avoid inadequate states. In a fourth case, behaviour must be modified in such a manner that this can be interpreted as a change of goals which the previous behaviour seemed to be aimed at. In a fifth case, the automaton will risk a series of inadequate states so as to display adequate behaviour at subsequent stages. In other words, the environment can be made more and more involved so that even the most "intelligent" automata will make mistakes. A case in point is the experiment described by Lewis Carrol in "Alice in

Wonderland" [2].

The conclusion suggests itself that intelligence can be manifested in an automaton's behaviour in an environment whose variation is limited but always present. For practical evaluation of intelligence, one must primarily be capable of apprising the diversity of the environment perceived by the automaton, the variety of its output responses, and of measuring the adequacy of the latter. The surrounding world seems, however, so variable and the behaviour we observe in it is so diversified, that it might appear impossible to find a general frame of comparison. It turns out, however, that Nature has found such a universal framework for describing the two.

Information entering and leaving the brain is presented in the same form - continually changing spatial-temporal configurations of nerve impulses. This form of presenting information has a very high capacity. The whole of the external world in its diversity and the interoceptive information, on the one hand, and the multitude of coordinated output responses, on the other hand, are encoded as multidimensional impulse texts. Behaviour of automata in a variable environment can be easily reduced to these texts.

2. Linguistic approach

Human and animal behaviour can be interpreted as a manifestation of the relationship of two texts constituted by the spatial-temporal structure of the nerve impulses. One of these texts will be referred to as afferent text (it is constituted by the external and internal receptive fields), the other one will be called efferent text (it is formed by the brain). The function of the brain in generating the behaviour consists in transforming one impulse text into the other. This transformation is conveniently interpreted as translation from one language (afferent) to the other (efferent) language. In this sense, instances of translation are: writing dictation, driving a car (visual afferentation is transformed into efferentation of the driving movements), and all other kinds of behaviour, including its complex forms (dialogue, creative processes, etc.). Intelligence of behaviour can be evaluated in terms of exactness and brevity of translation. Evidently, mistakes in translation are identical to inadequate behavioural responses. The behaviour of an animal in a strange environment made up of elements of a priorly known environment resembles translation of an uknown text consisting of words that are known. The flexibility and adequacy of the behavioural responses in this case are equivalent to coherent and error-free language translation.

What is the accuracy of evaluation to be expected of this linguistic approach ? Each customer of a translation service knows by experience that the knowledge of a foreign language is no warranty of correct translation. This confirms Ashby's words on the choice of criterion, although his situation is better compared to "editing".

The channel of the afferent text in man contains $4 . 3 \cdot 10^6$ nerve fibres (posterior radicles and the intercranical afferent nerves, including optical and acoustic ones). The output channel contains $3 . 4 \cdot 10^5$ fibres [3]. The transmissive capacity of the afferent channel in man is greater than that of contemporary technical facilities. That superiority is not too great, however: present-day mass media — radio, cinema, TV - transmit almost as much information. As to the efferent channel, it transfers one-tenth of the afferent channel capacity, judging by the above figures (or even one-twentieth, if the spinal cord is not counted). Consequently, present-day media are well capable of recording and reproducing the output responses to the slightest shades and details.

The problem of artificial intelligence does not boil down to mere recording and reproduction of the above amount of information, but lies with providing associative connection between input and output information, accomplishing adequate "linguistic" translation of the afferent impulse text, not known in advance, into the efferent text. Human and animal brains easily produce this associative connection, even though with varying adequacy. The task remains a problem for the machine, for instance, in the case of automatic translation. Technology gives facilities for recording the huge amounts of information mentioned above only as a function of time — using various time-based scannings. To cope with this task, however, it is necessary to record, reproduce and, obviously, lay out the efferent text on the medium as a function of the informational structure of the afferent text.

Thus, to implement behaviour simulating language translation it is necessary to provide associative connection between the above information amounts. The recording-reproduction methods known at present are unfit for the task.

This gives rise to the need of using a new type of scanning for recording, that we shall refer to as informational scanning.

Note that for technical purposes it is not necessary to reproduce an amount of information comparable to that reproduced by man. To begin with, it would be sufficient to reproduce the associative behaviour of, say, an insect.

This involves relatively small information amounts, all the more so as only efferent texts must be recorded. The memory device for the afferent texts is the environment itself. In this framework, the sensory organs are but devices for reading out and encoding information external to the brain. This information must be used to perform the translation (more exactly, synthesis) of the adequate efferent text. Afferentation, so to speak, controls the addressing system of the memory, performs a kind of scanning by which the appropriate areas of the efferent text are selected. Assessing the adequacy of behaviour for the case of the associative recording of two informational texts, one can say that the intelligence of the behaviour is equivalent to "word-for-word" translation. This problem statement is analogous to the holographic method of information recording, where the informational "scanning" is to an extent affected by the source of the reference coherent signal [4]. The computer-controlled adaptive robots operate according to the same pattern: the system of gathering information on external and internal environment builds the afferent text, which is "translated" by the program into the efferent text that controls the robot's movements. Let us note incidentally that all systems under consideration are closed in the sense that the movements of the device and the changes it introduces into the external environment are reflected in the structure of the afferent text.

In comparing automata with living organisms, one is struck by the involved way in which the afferent-to-efferent translation is made by the computer. The enormous high speed of the computer is usually wasted on dictionary lookup. The responses to environmental signals become rather slow, and the information capacity of the efferent flow is very low as compared with the potentials of the up-to-date data recording-reproduction facilities. This leads to the idea of developing a "direct" method. In this connection, an attempt has been undertaken at the Polytechnical Institute in Leningrad of devising the principles of an associative memory with addressing to be controlled by the afferent information flow (informational scanning) and to determine the lay-out of efferent information on a medium [5]. A memory of that kind would be convenient for recording the intelligent behaviour in a particular, possibly varying in a rather involved way, environment; then a series of experiments in scanning that behaviour using a different, but to an extent similar, afferentation would be conducted. We shall discuss below the intelligence potentialities of such a **memory**, where the efferent text once recorded is used as "quotations" for the synthesis of many other efferent texts in application to various afferent situations.

3. Artificial intelligence at the level of word-for-word translation

The concept of informational scanning developed at the Polytechnical Institute (Leningrad) is illustrated by the figure, using the following notation:

1 — modulatable light source for recording (during the read-out it is switched on constantly),

2 — transparent elastic medium capable of superposition of excitations,

3 — set of exciters transforming the afferent text X into the dynamics of the interference patterns of medium 2,

4 — source of afferent text (for convenience of discussion it is represented by a multi-track tape-recorder whose tape can be cut into "quotations" and then pieced together into other afferent texts thus synthesized),

5 — threshold converter of dynamic interference patterns, which identifies in these the excess points - where excitation surpasses the threshold,

6 — photo plate (the carrier of efferent information); the mosaic of black and white spots on it determines whether the light flux from this or that excess passes through it, or not, and

7 — light detector which reproduces the efferent text by recording the flowing of light through the photo carrier 6.

The impulse afferentation X is transformed by the set of exciters 3 into interference patterns which exist in the form of elastic interactions in the translucent optically active medium 2.

This transformation of the afferent text is linear, i.e. there is always one-one correspondence between the interference pattern and the gliding fragment of the afferent text. Interference patterns are used for addressing the light beam of the source 1 to the information photo carrier 6. For that they are subjected to threshold conversion which singles out from the interference patterns only the superliminal points - excesses. There is a scan for the excesses on the surface of photo carrier 6, the scanning trajectory is determined by the spatial-temporal structure of the afferent text. On these trajectories any other - in particular, scanning - information U can be recorded which modulates the brightness of light source 1.

Subsequently, when the afferent text X is reproduced with the source 1 switched on constantly, the information U written on the photo carrier 6 can be reproduced by the light detector 7 during the recording of the light flux penetrating through the photo carrier. In this way, any efferent text Y can be shaped and reproduced.

The informational scanning outlined above possesses certain features that make it interesting for pattern recognition, associative behaviour simulation and research into the intelligence of this behaviour.

It has been established [6] that, in the dynamics of the interference patterns, the excesses change their locations in a leap-like manner (this is a consequence of the mapping of the multidimensional information process X into the plane). The location of each excess in that case is determined by the spatial-temporal structure of the gliding fragment of the afferent text X. At the same time, limited (minor) variations of the spatial-temporal structure of this text cannot shift an excess from its location. In an extreme case, an excess may disappear or appear in another place (which is precisely the leap-like character). Hence follows the essential conclusion as to the detecting properties of the memory: an ample set of insignificantly differing texts X is reproduced by one and the same efferent text Y. In another word, minor changes of the external or internal environment do not lead to a change of the standard behaviour of the automaton. This is the first type of response to a change of environment (pattern recognition, adaptation to noise).

If tape 4 is cut into "quotations" and a new afferent text is synthesized out of these, at the output 4 a new efferent text will be synthesized out of fragments of the text previously recorded on plate 6 with a corresponding change of sequence. The behaviour of the automaton will be equivalent to word-for-word language translation, except for one distinctive feature: there is no dictionary, for there is no a priori segmentation of the afferent text. The initial text is segmented automatically each time in a new fashion in keeping with the new configuration of the afferent quotations. To provide for arbitrary segmentation it must be admitted that a "cut-out piece" or its parts are not excluded from subsequent use. A dictionary-like ordering of the initial record (during the learnig) is superfluous. On the contrary, the record having the coherent form of natural behaviour (translation) is more efficient, for a "word group" will be translated as an integral "quotation". It is also justified because otherwise the quotations in the efferent text would be separated by blanks at the new junctures of the afferent quotations. The blanks will appear because the junctures constitute, so to speak, quite new text fragments whose translation has not been written.

The economical recording of the "dictionary" is noteworthy. Only the second (efferent) part is recorded. The first (afferent) part is not recorded at all, although this input part of the "dictionary" is larger by many orders in view of the dectective properties of the memory described. Moreover, repeated recordings of situations, where the same X and U are for a second time entered into the memory, do not result in extra spending of carrier 6, as they are addressed to the same trajectories and repeat on them the old traces. Also, thanks to the dectective properties of the memory, the text X can vary to a certain extent.

Certainly, the intelligence of the "word-for-word translation" behaviour is not very high. However, it has a difference to its advantage for the case of computer application - the possibility of quick manipulation of quite ample amounts of afferent information. There is no "bane of multidimensionality" which invalidates modern computers in respect of certin tasks. Information capacity of efferent information can be also essentially expanded. It would be appropriate in this context to compare the experiments in computer-based synthesis of human speech (which are very far from perfection) with the reproduction of speech by a tape-recorder. The operation of the associative memory described above is more similar to the latter case, so that no problems of reproducing the individual features or shades of voice have to be faced.

The difference is that instead of the tape-moving device another type of scanning is used, which affords selective read-out of the excerpts of the efferent text recorded to be carried out in any sequence depending on the changing afferent situation. Let us emphasize that in the case of active behaviour the efferent text and the afferent text are closed and independent, which determines the coherence of behaviour and its intelligence. Besides, as was demonstrated in [6], the efferent text can simultaneously serve for the afferent one, in which case the memory has the properties of an extrapolator with associative detective access. Such is the type of response to environmental changes (associativity at the level of re-arrangement of fragments and extrapolation). Restrictions are imposed by the need of facing "quotations" long enough. Otherwise sensible information losses are incurred at the "junctures" of the excerpts.

It must be noted in conclusion that the behaviour of the "word-for--word translation" type is far from always sufficient for the requirements of a designer of an intelligent automaton, and in particular, a robot. It is evident that in the general case this kind of translation may be regarded solely as an element of a more complex hierarchical system, which must also incorporate an electronic computer at certain levels.

An automaton is not always capable of judging about the adequacy or inadequacy of its behaviour, in which it is no better than man. Therefore, a more perfect intelligence system must be used for the evaluation of more information, which is obtained by purposeful experimentation (including simulation, consultations, etc.). As the result, a learning automaton will be able to upgrade its intelligence. This does not mean, of course, that a learning automaton always has a superior intelligence. A non-learning automaton may display a more adequate behaviour owing to the abundant "genetic" information it has.

REFERENCES

[1] Ashby, W."What is intelligent automata„ "Cybernetika ogidaemaja, cybernetika neogidannaja . Nauka, M., 1968

[2] Carrol, L."Alice in Wonderland„

[3] Blinkov, S.M., Glezer, I.I."Human brain in figures and tables„ Meditsina,
Leningrad, 1964. (In Russian).

[4] P.R. Westlake,"The possibility of neural holographic processes within the
brain„ "Kybernetik„. Band YII, Heft 4, Sept., 1970

[5] Radchenko, A.N., Gol'din, V.E."A study of a model of associative behaviour„
"Avtomatika i Vychislitel'naya Tekhnika „ 6, 1969. (In Russian)

[6] Radchenko, A.N."Analytical studies of the neurophysiological processes of
retention and reproduction of information on a conceptual model„
In: "Kiberneticheskie Aspekty Izucheniaya Mozga „ Nauka Publish-
ers, Moscow, 1970. (In Russian).

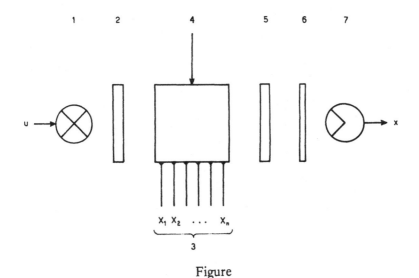

Figure

DESIGN STUDIES ON INDUSTRIAL ROBOTS

Meredith W. THRING, Professor
Queen Mary College (University of London)
Department of Mechanical Engineering
London, England.

(*)

Summary

      The Robot (a machine pre-programmed to do any one of a range of complex manipulative tasks) is in a very early stage of development because all the ones in industry have no sensory adaptability and cannot do tasks involving moving themselves about the factory. Most of the work on giving them touch, force and visual senses requires a large computer to enable them to adapt their movements to sense impressions in the pre-instructed way. Our work is concerned with developing sensory systems which can be readily adapted to different complex situations using only small digital or analogue computers or devices like un-selector switches. We have built a device that seeks out, picks up and carries away objects of widely varying size and shape placed at random on a table, a hand that grasps an object of varying size with a light or a heavy grasp and we are designing a self-propelling carriage in which a scale plan of the permissible paths in any factory can be placed and it will follow these paths by dead reckoning, avoid unexpected objects and locate itself accurately by tactile sensors.

(*) All the figures quoted in the text are at the end of the lecture

1. Why build robots ?

The Engineer who works on the design of robots is faced with exactly the same dilemma as the Engineer who works on motor cars or the production of DDT. Rightly used his product can be of great benefit to ordinary people, but wrongly used it can cause immense harm. In the case of robots the right use is to free humans in an industrial society from boring, repetitive work which requires no proper training and gives no satisfaction to the person who does it and also to free them from working in dangerous, uncomfortable or inaccessible places. The grave danger is that robots will be used in such a way that they increase unemployement especially of people who do not have their strength in the intellectual brain. It is my prediction that, unless we change to a policy of rearranging the paid work of society so that there is an interesting worthwhile job for everyone, robots will be the last straw that breaks the camel's back by causing unemployment to rise to intolerable levels. Norbert Wiener, the founder of cybernetics, made a similar prediction 25 years ago. If, however, we ration work by shortening hours, and we increase human service activities (e.g. 10 pupils to every teacher, adequate nursing facilities, proper care for old people) and use enough labour on the land to get high crops on a perpetual basis, then the robots will be an unmixed blessing.

In this paper I shall discuss the possibilities of robots from this point of view, that of the benefits they can give to ordinary people if they are used as a part of the Creative Society, i.e. as slaves to serve our needs. Just as the glory of Athens, was based on human slaves so we can have a society in which everyone in the world has full opportunities to realize their creative possibilities if the robot slaves enable the humans to earn their living with a fraction of their lifes energies and without drudgery.

At the present time there are two kinds of factory; those where a very large tonnage of a single product or millions of identical objects are made, and the factory which changes its product relatively frequently. In the former, which includes steelworks, glass works, power stations, oil refineries and mass production assembly lines, the factory is laid out so that the product goes through mechanically as it is gradually processed and the job of the humans is either supervisory or some completely repetitive task on each unit as it passes on the assembly line. It is becoming recognized that this last type of work is so sub human that it is actually more economical to scrap the production line and give a team of humans the job of being completely responsible for the construction of an assembly so that they can take pride in the quality of their work. The two Swedish car firms Volvo and Saab are reporting good results on these lines. This is essentially the method proposed by Willian Morris 100 years ago – scrap the machines and return to hand craftmanship. It would be nice if we could do this for all production work but unfortunately the world has now so many people and the benefits of mass production are sufficient that we could not do it without condemning the underdeveloped countries to a perpetually depressed standard of living and substantially reducing the proportion of people with valuable machines in the developed countries. We cannot therefore put the clock back but we have to try and move forward to a point where everyone in the world has the real benefits of technology without the disadvantages of pollution and exhaustion of raw materials (especially fuel).

This means that we have to consider a world in which enough consumer and capital goods are available for everyone (7000 million people in the XXI Century) to have good food, housing, clothes, health, education, freedom to travel and to walk in open country and opportunity and materials to develop his or her creative talents to the full whatever they are. It thus requires the development by engineers of designs of machines for public and private transport, building, agriculture, preventive and remedial medicine, leisure activities of all kinds, communication and education, which can be produced in vast numbers and yet which make minimum use of row materials, by being very long lasting, economical of fuel, and by recycling of all waste products. Such machines must be made by mass production methods so good that the quality of all work and the reliability and safety of the product is guaranteed at a very high level.

I think this argument does lead one inevitably to the conclusion that robots will have a major part to play in the factories of the Creative Society and that humans will be the masters of the robots, doing supervision, repair and maintenance. No human can ever satisfy all his creative activity in his paid work because no work, even that of an artist or singer, ever gives a fully balanced use of all three brains (emotional, intellectual and physical or heart, head and hands) so in the creative society the shorter hours of work given to everyone by the robots will free them to put time and energy into a freely chosen creative hobby done not for money but to give them opportunity to stretch all their talents to the limit.

Thus we can envisage that there will be assembly line factories for mass production in the Creative Society, but instead of a row of humans doing each operation on the line, a row of robots will do them and will check each operation and report any failure to the human maintenance staff. The use of the Unimate for welding in a GM car assembly line is an already existing step in this direction.

There will probably still also exist the second type of factory in which at present humans unload incoming lorries bringing raw materials or sub-assemblies, put them in a large store, take them from the store to a variety of machines, feed the machines, assemble the mechanical parts, package the product and load it onto outgoing lorries. All these tasks can be done by robots and the setting up of the numerically controlled machines and the routine control of all the robots will be done by a central computer with which they will be in constant communication by closed loop radio so that for example they can be given paths which avoid collisions with one another.

In addition to the factories robots will probably be extensively used in mining, undersea work and for routine cleaning and other operations in the ordinary home. Mining of coal and minerals represent the clearest example of work which is dangerous, uncomfortable and inaccessible so that there are obviously immense advantages in a system which enables us to win solids without men going underground exactly as we can win oil and natural gas. We already know how to pump crushed coal or any solid in a stream of water as oil can be pumped so the basic problem is to develop a machine which will find the seam and then work through it crushing the material in front of it as it burrows through dragging pipes and cables behind it. In the case of coal or other concentrated seam minerals it will pump the whole of its spoil to the surface, whereas for very dilute precious minerals like gold, uranium or diamonds, it may sift the required mineral from the crushed rock and leave the latter in the seam. Probably the actual miner will be a true robot, that is it will be controlled by a computer on the surface so that it will seek out and follow the seam automatically using chemical analysis density (by γ-rays), colour observations, hardness measurement, calorific value (by a small combustion furnace using oxygen passed down in a pipe) or some other measurement to detect the edges of the seam. However it will probably be necessary to have a telechiric machine which can be sent down to repair the robot mole when it breaks down with full communication to a human operator. Similarly all work under the sea will be done by telechiric machines or robots controlled from the surface. Probably it will not be necessary to have drilling platforms above the sea surface since the derrick will be anchored on the bottom of the sea and all the drilling operations will be done by machines controlled from the shore.

As far as the home is concerned I am convinced that every human family likes to express its individuality most in the home and thus that any homes with built in automation or ant heap aggregation are inhumane. On the other hand it is a pity to give good education to people and then expect them to spend a large part of their lives doing utterly routine household chores that have to be repeated every day or every week or even like spring cleaning once a year. Thus we can say that a domestic robot will be developed which can be taught where the furniture belongs and how to do all the routine cleaning, scrubbing, dusting, tidying, bed making, laying tables, loading and unloading the dishwasher of clotheswasher, preparing vegetables (but most people find cooking a creative task worthy of humans).

2. Definitions and Functions of Robots and other Humanoid Machines

A robot is essentially a computer with arms, legs and senses. It is a trainable moron which can have great strength, reliability, tirelessness, docility and memory but only limited manual skill. I would define a robot as follows:

A machine which can be pre-programmed by a human master to carry out any one of a variety of complex manipulative operations (often involving self movement on a complex path) and to vary the details of its movements in accordance with its own sense impressions of variations in its surroundings in a manner to which it has been pre-instructed by its human master.

Fig. 1 shows a block diagram of the elements of a complete robot in its relations with its surroundings. It can be simplified by the elimination of the ability to move its own sensors, or even (as in the senseless non-adaptive robots in use at present) by the elimination of the senses; the feedback sensation of its own position or orientation or of the position of its hand may be eliminated in the case of positive control.

The essential components of a complete robot are thus as follows:—

1.   One or more hands and arms.
2.   Self propulsion and self steering.
3.   Self contained power and control systems for 1 and 2.
4.   A limited computer with memory for sequences of instructions and for interpreting sense impressions to make instructed deci-sions. Possibly this computer can be in real time communication with a larger central computer.
5.   Senses — analogue or digital observations of some of the following
     a) Touch roughness, hardness, position, weight thermal conductivity temperature, proximity shape and size by gauging comparison.
     b) Sight: Shape, colour, distance, size.
     c) Smell: gas phase chromatography.

d) Position of its own links, also force on them.

e) Hearing: e.g. to respond to the shouted order "stop".

f) Other special physical or chemical measurements.

The second category of humanoid machines is the telechiric machine; the name was coined in the USA from the greek words for distance and a hand. A telechiric machine is a machine which can copy sufficient of the movements of one or a pair of human hands to do handling operations at a distance from the operator and given sufficient sensory feedback to the human so that he can do the movements as skilfully as if he was present. The hands of the machine may be much larger and more powerful than those of the human operator, the same size, or much smaller and capable of the corresponding scaled down accuracy for very fine work, e.g. in surgery, dissection or very fine machine or electronic circuit construction. The machine has no memory and is not capable of being instructed to make its own decisions so that it is entirely dependent on the brain of the human operator all the time it is working. The basic engineering problems in the design of such machines are the mechanism to provide accurate positional correspondence between the operator's hands (and legs also in some cases) and the hands of the machine independent of load, with some controlled force feedback ratio so that the operator knows when the hand comes against a resistance or picks up a weight too heavy for it, and with visual feedback. In addition to the six controlled movements (3 translational, 3 rotational) which must necessarily be provided for each hand, there must be at least one for grasping unless the hands are purely sockets with which tools can be picked up and held. The possibility of doing skilled work in a dangerous situation with the human in a safe place has already been applied to nuclear reactors, undersea work and the Russian moon machines and will probably be extended to the extinction of fires (e.g. oil wells, crashed aircraft, buildings), the rescue of humans from a disaster area and to repairs of broken down machinery in inaccessible places as mentioned above for mining. The great strength potential of such machines combined with the human skill (e.g. to pick up a box by grasping its sides between finger and thumb without crushing it) can be applied for aircraft and ship loading and accurate and rapid placing of heavy machinery. The use of telechiric micro hands for surgery has not only the possibility of increased precision but also the advantages that (1) the surgeon can control one pair of hands exactly into place and then lock them and move his hands to another pair of controls, (2) the surgeon can be outside the operating theatre so that sterilisation is much easier, (3) he can control several TV cameras to see from different points on several screens at once and one can operate from a flexible fibre optic bundle to see round the back of an organ or inside the body, (4) the design of the hands and tools can be made more compact for a given strength than a human hand.

The third group of humanoid machines are the mechanical limbs. These may be arms, hands or legs but they are mechanical artefacts attached to a human body close to the normal position for such a limb and operated by the human. They can be subdivided into two sub-groups. The first is the substitute limb for a human who has lost the limb or been born without it. A great deal of valuable work has been done on artificial hands and arms powered by electric batteries, compressed gas or mechanically from other muscles, and controlled by some bodily movement. Unpowered artificial legs have been used for many years but are not much use unless the user has the three thigh rotations relatively unimpaired. I have started a program to try to develop powered artificial legs controlled by moving the arms but this involves much more difficult problems than the powered artificial arms because much more power is needed and the control for dynamic balance in walking and stair climbing is very complex. Waseda University in Japan has produced a series of powered walking legs which contribute a great deal to the solution of this problem.

The other sub-group of mechanical limbs is the exoskeleton – Fig. 2 shows an unpowered exoskeleton leg pair which we have constructed to help someone who has arthritis to stand and walk with less pain because the weight of the person rests on the bicycle saddle and is taken through the exoskeleton to the metal soles under the user's real shoes. Studies have been reported by Ralph Mosher of General Electric Pittsburgh on the possibility of a man walking inside an exoskeleton of giant strength with its own power pack so that its legs move with his legs and its arms can lift vast weights when he moves his arms. Its limbs have to be essentially the same length as his limbs and to be maintained close to his limbs as they move by the force-amplifying motors in the joints of the machine.

3. Limitations of Robots

A robot is an artefact, designed and made by men and I believe that man will never be able to build into his artefacts certain abilities which he himself possesses as a result of the biological processes by which he is formed. One cannot prove the following postulates of impotence any more than one can prove the first or second law of thermodynamics, but sufficient failures to overcome them will put them on a similar basis.

The first postulate corresponds closely to the second law of thermodynamics stated in the form that all systems run down to a "heat death" unless consciousness (Maxwells demon) intervenes and to the law of communication that a mechanical system can only make a message more garbled. This postulate is:

Postulate 1. A robot can never do any task more sophisticated than it has been instructed by a human to do, e.g. it can sort things into categories that have been given to it but it cannot by itself decide to create a new category. It can carry out its instructed task more patiently, accurately and tirelessly than any human can do, but if it makes errors, these will be random and not improvements or self correcting (unless it has been programmed to check by a different process).

> **Corollary 1.** Robots left unattended by humans will gradually lose their working effectiveness and eventually cease to work altogether because however much the ability to repair one anothers breakdowns is programmed into them, unforeseen breakdowns will inevitably occur which they are unprepared to diagnose and correct.

> **Corollary 2.** If robots are made which can be programmed to build copies of themselves, then by the third or fourth generation so many faults will have arisen that they are useless. Incidentally if we did not have bi-sexual reproduction and laws against incest the same fate would probably be ours.

> The second postulate of robot inferiority to humans may be called the aesthetic law.

> **Postulate 2.** No robot or computer can ever be built by humans which will have true human emotions or feelings. Thus whereas a human can be educated and learn to use all three brains –

Intellectual (dealing with logical abstractions, mathematics, scientific theories)

Physical (the trained skill of the hands, eyes and body, e.g. in running, using tools or driving a car)

Emotional (concerned with feelings like happiness, loyalty, love, artistic appreciation, quality of life, value judgements e.g. people matter infinitely more than machines)

a robot is necessarily limited to the last two.

> It is of course true that a robot can be made to produce all the exterior manifestations of emotions as does a sound film but that certainly does not mean that the mechanism itself experiences the emotions. This postulate can certainly never be proved and indeed in most science fiction (e.g. 2001-The film based on a story by Arthur Clark) it is assumed to be false. However all matters dealing with emotions

are essentially subjective and so beyong the realms of science and technology. I believe that all attempts to build a robot that in itself experiences human feelings, even the jealousy and vanity of HAL, will fail.

Corollary 1. Since the emotional brain is essential for all really original creative work (such as scientific hypothesis formation, recognizing a new phenomenon, painting a great picture, composing a new symphony or inventing a new type of machine) it follows that robots will always be barred from all these activities.

Corollary 2. The only value judgements robot can have are those built into it by its human maker or programmer. Designers of robots have thus a special responsibility to build in respect for all human life, for the environment and for human feelings. We have to design robots as perfect slaves to humans, but we need have no shame in doing so since their lack of an emotional brain makes them our things.

4. Design Principles for Robots

1.   The appearance of robots will be functional and they will not look in any way like human beings. Plays about robots naturally make them look like humans because they are human actors. However the materials used in engineering have a wider range of possibilities and especially strength/weight ratio than the materials of the human body, the engineer can choose arrangements that are much more stable and protected against injury than the human. He can also make joints that rotate continuously although this may land him in problems for communicating control signals reliably. He never has to design a robot with anything like the fantastic range of learning possibilities for movements and skills that the human has, so he can simplify very much. On the other hand he cannot construct mechanisms nearly as complex and automatic controls or brain controls nearly as subtle as the human ones of muscles bathed in a blood stream bringing them their fuel and oxygen, the temperature and chemical controls, and the nervous system the brain with its interconnections and multiple linkages, ability to learn to recognize word meanings, local accents, faces of friends, smells. Instead of using craft tools like hammer and chisel the robot will carry around a series of power tools operated from its own battery or power source which it plugs into its arm in place of its hand. It will probably consist of a low truck-like base carrying all the heavy machinery inside it and running on a centipede track (fig. 3) or on four rimless wheels (fig. 4) which enable it to move in any situation where human legs suffice (rock climbing which

needs the arms as well will require a special robot if it has to be done e.g. for rescue work).The brain would be in the base but a vertical column may carry two arms or one arm with two hands.

2.   Independent control mechanisms for each movement are elaborate and difficult to construct and therefore robots must be designed with as few of them as possible. Examples in our work are (1) the hand (fig. 5) which grasps the same range of shapes and sizes of object as the human hand from a niddle too a cricket ball but uses only one independently controlled movement instead of the 20 separate movements of the 5 digits of our hand. (2) The stairclimbing carriage (fig. 4) which only requires one controlled drive as opposed to at least two with the simplest possible leg pair (fig. 6) and to six in each leg used by a human.

3.   The computer brain for a robot must use the simplest possible system for programming, decision making from combination of sense impressions and memory. It is not right because digital computers have been developed so fast to assume that digital computing is necessarily the whole answer to the robot brain. It is almost certain to contain specially designed elements of a proportional or hybrid character. Our own table clearing robot (fig. 7) finds the object by traversing along the table until a light beam reflected from a fixed reflector along the back of the table is interrupted by the object and than an arm comes out to find the object across the table. When a lightly spring contactor switch informs the hand that it has found the object the arm stops moving across and the other side of the hand swings down and closes on the object which may vary in size from a large dinner plate to an egg cup. When it senses a firm grasp on the object, by a heavily sprung micro-switch in the same part of the hand that sensed the object it stops closing, grasping the object between the two parts of the hand which are faced with high friction plastic (sticky) backed by 5 mm of foam rubber to give a good grip. It then lifts the object off the table and the arm withdraws to its closed position so that it can release the object on to a round disc shaped tray which then rotates so that the object is carried away from the hand and it is guided from the tray by a fixed vane at the back. The robot then continues to move along the table until it comes to the next object, when it repeats the procedure. It is not yet sophisticated enough to deal with two objects in line with respect to the light beam. The point of describing this robot in detail is that while it can adjust its cycle of movements according to the three sense impressions of position of the object along the table, across the table and to the size of the objects its "brain" is no more complicated than a single second hand uniselector switch from the automatic telephone and its senses are a photo electric

cell and two micro switches with springs of different strength. In an earlier study of a mechanical rat in a maze (fig. 8) the rat could be made to explore all the eight paths sequentially returning to the standard position at the start (top of fig. 8) but to change its sequence of operations to one in which it repeated a particular path as long as a "piece of cheese" (a projecting object when it came to the end of the path) was placed in the path. Here again the uniselector switch provided all the brain needed since it could repeat a sequence or pass on to the next according to which of two micro switches was operated at the end of a path.

We can consider as a more advanced example the robot that will carry objects about a factory from one place where it is instructed to find them to another where it is instructed to place them. The first information it has to be given is the possible paths in the factory that it may take. This information can clearly be laboriously programmed in to the memory of a digital computer, but a much simpler way will be to use a simple scale map of the factory as we did in our robot fire-fighter (fig. 9). In this case the robot can follow its real path in the factory by traversing the map using a compass to sense the direction of its movement and counting wheel revolutions to estimate the distance. Moreover since dead reckoning always leads to cumulative errors it will be necessary for the robot to make direct observations of its position in the factory by sight or touch sensors using landmarks. These landmarks can be marked on the map and so it can correct its position on the map and by observing two landmarks or a horizontal line on one of known orientation it can correct its orientation. If a human controller wishes to program it for another factory or to block out a path which is now forbidden he can change the map with a pen directly so that communication of this type of information to the robot is greatly simplified and much less liable to error. It is clear that the amount of electronic computer elements is greatly reduced and there is no need to have the high speed of a digital computer because the map follower can easily keep up with the real movements of the robot.

4.   The designer must build in safety devices to meet any dangerous mis-functionings that he can foresee. For example the idea of the robot going berserk and endangering humans can be met by a device which switches off its power when any human voice calls "stop".

5.   The power source of a robot must be compatible with the environment within which it is to work. This, at present, rules out the most portable power pack – the internal combustion engine – if the robot is to operate indoors, although it may be possible to develop an engine such as a free-opposed piston

engine running as an air compressor with methyl alcohol or compressed methane as the fuel which reduces the noise and only emits CO_2 and H_2O. One must also develop systems in which very light, simple actuators e.g., compressed air cylinders and rotary actuators operate all the movements, or else in which a single motor operates all the movements by a system of electromagnetic clutches (fig. 10).

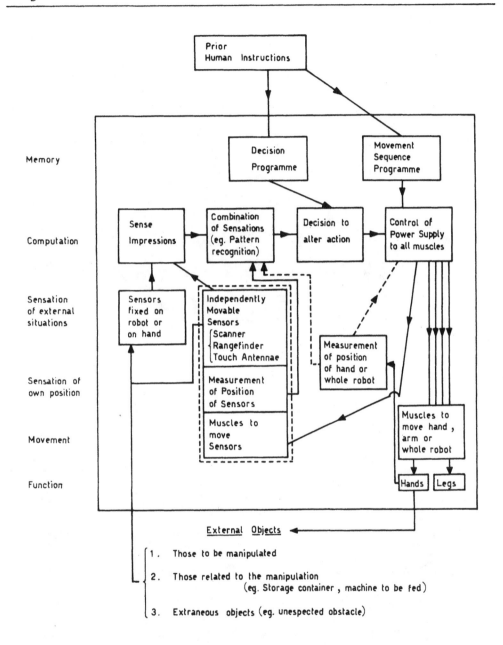

Fig. 1   Block diagram of control system of generalized robot

Fig. 3 Centipede

Fig. 2 Unpowered exoskeleton leg pair

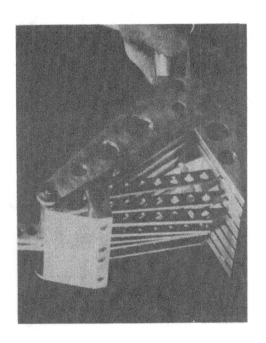

Fig. 5   Mechanical hand

Fig. 4   Stair climber

Fig. 7   Table clearing robot

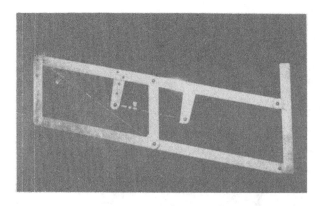

Fig. 6   Model for artificial leg

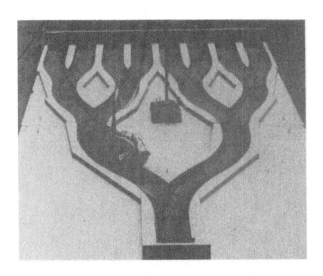

Fig. 8   Mechanical rat in a maze

Fig. 9   Robot firefighter

Fig. 10 Use of electromagnetic clutches for arm actuation

6. APPLICATIONS

IMPACTS OF TELEMATION ON MODERN SOCIETY

Arthur D. ALEXANDER, III (*)

(**)

Summary

A national survey study of state-of-the-art technology, major civil sector user needs and recommendations for Federal initiatives in the field of teleoperators, robotics and remote systems is reviewed. Impending developments in the application of telemation to remote emergency medical care and to remote mining operations are speculated upon.

(*) Research Scientist, NASA Advanced Concepts and Missions Division, OAST, Moffett Field, California 94035.

(**) All figures quoted in the text are at the end of the lecture

INTRODUCTION

The Office of Exploratory Research and Assessment of the National Science Foundation and the National Aeronautics and Space Administration requested me, in late 1971, to undertake a ninety day study of Teleoperators, Robotics, and Remote Systems Technology in the United States. The purpose of this study was to survey state-of-the-art technology, in this field, determine major user needs in medicine, mining and oceanography, and suggest initiatives where federal R&D funding would most significantly impact the application of this technology to the alleviation of explicit national social problems.

Following a review of the findings of this study commencing with user needs, I shall speculate on impending developments in the application of telemation to remote emergency medical care and remote mining systems. These speculations are based upon extrapolations of work in progress.

USER NEEDS

Medical

Despite the present availability of "space-age" telemation technology , there has been only minimal transfer of this know-how to the medical diagnosis, treatment or rehabilitation of the sick or handicapped.

Every year, more than 1 of every 1500 Americans dies because of trauma shock following an accident or medical emergency, much of the trauma occurring after the main event prior to effective medical care. Over the years, almost 1 American in 10 has become physically disabled to some degree, a large fraction of the permanent disablement attributable to delays and unskilled handling before receipt of effective treatment. In much less affluent nations than the USA, except for emergencies happening under the most favored conditions, the victims, often suffer worst possible outcomes of their afflictions, Table 1 indicates that nearly three million of these disabled Americans could be helped significantly with current orthotic or prosthetic rehabilitation technology. Many of the remaining sixteen million are rehabilitable to a degree that would allow them to become productive individuals. Their rehabilitation would release trained therapists and aides to work with the more seriously disabled individuals who may not be rehabilitable with current technology.

The overall goal of rehabilitation engineering is to improve the self-sufficiency of the patient whose mobility is restricted by illness or disability.

The ultimate test of the usefulness of any devices which assist patients is the degree to which the patients can become useful citizens, capable of using their talents in gainful employment. It is our conviction that the generation of manipulators and associated devices currently under development will make it increasingly possible for disabled individuals to perform useful work. The potential positive economic effects upon GNP would appear to be significant.

The use of remote manipulators for surgical care provides an enormous benefit in avoiding tissue trauma. One of the greatest sources of surgical morbidity at present is the need to create additional tissue trauma in order to achieve adequate exposure to accomplish the corrective manoeuvers necessary. The technical skills of the master surgeon are marked by his ability to achieve exposure, accomplish the job, and withdraw with a minimum of lost time, blood, and exposure to the internal environment.

Current medical institutional practice routinely employs highly specialized manipulators for certain surgical procedures. These include instruments for visualizing and operating within the interior of the stomach, bladder, intestinal and urinary tracts, the female reproductive system (tubal ligation), the heart (pacemaker implantation) and blood vessels, the inner ear and brain. These instruments are principally modified conventional surgical instruments designed for insertion and use at the operative site so as to minimize surrounding tissue trauma often associated with major surgery. The effectiveness of these procedures has reduced many major operations to the category of relatively routine minor surgery; requiring but minutes to perform, minimal anaesthesia, and only a few hours for recovery. Cost and risk to the patient are thereby significantly reduced.

Mining

Mineral economists have often noted the stark fact that the mineral needs of this country are increasing at a phenomenal rate. During the last thrity years, the United States alone consumed more minerals than had all preceding generations in history. We use more minerals than we produce. For about half of our sixty most-important minerals, we must rely in whole or in part on foreign sources of supply. This contributes to our balance of payments problem and weakens our national and economic security.

If we are to reverse this course we must expand domestic mine productivity and seek new "captive" sources in the future. In order to do this, we must provide the technological base necessary to permit the efficient extraction of our nation's mineral requirements while maintaining safety and environmental quality at an acceptable level.

The extraction, as well as the development stages of mining, will require advanced mining techniques to cope with the harsher environments imposed by mining at increased depths. Completely automated systems of excavation and mineral extraction are needed to permit workers to perform their tasks away from areas of poor ground or environmental conditions while simultaneously obtaining economical mineral extraction. Required is the implementation of remote control and sensing technology to provide a mechanism for discriminating between valuable and waste material and a means of communicating this information to remote locations where feedback control mechanisms to direct the mining system can be employed.

Oceanographic (*)

Undersea Application. Most of the ocean floor is at least two miles deep. Since divers rarely work for prolonged periods of time at dephts greater than two hundred meters, it is imperative that methods and equipment be developed that will enable man to exploit the commercial, scientific and military potential of the ocean floors. Substantial petroleum reserves under the deeper portions of the continental shelves have given commercial impetus to undersea technology.

Although many operational problems of inner and outer space are similar (viz., the necessity of firmly anchoring the teleoperator vehicle to the target), the environments have radically different effects on teleoperator design. The undersea teleoperator is surrounded by a good heat sink, but one that is extremely corrosive and laden with silt and biological agents. The tremendous pressures at great depths preclude the common mechanical master-slave linkages between the control and actuator spaces. The sensor problem is also different. Instead of the bright sunlight of orbital space, there is such darkness that an operator cannot see a manipulator hand which is only a few feet in front of his viewpoint.

Both in outer space and under the sea men may have to identify, build, maintain, repair, recover, or destroy some object. These activities require cleaning, bolting, cutting, welding, replacing parts, etc. — just the things men's hands do to terrestrial, dry-land equipment.

(*) Excerpted from "Teleoperators and Human Augmentation", NASA SP-5047, by E. G. Johnsen and W.R. Corliss, 1967.

In all oceanographic applications, the indifference of teleoperators to time, fatigue, and the hostile properties of the deep-sea environment is of fundamental economic importance. Keeping ships at sea and divers on the bottom are costly operations. The advantages of around-the-clock oceanographic teleoperators are obvious, particularly for commercial and military applications.

TELEOPERATOR SYSTEMS AND SUBSYSTEMS

At this time I would like to describe the various major remote teleoperator systems, teleoperator control/communications subsystems and manipulator devices surveyed. These sytems and subsystems represent only a fraction of the composite work in the teleoperator field, yet are characteristic of the present state-of-the-art technology. (Chart 1)

Chart 1
TELEOPERATOR SYSTEMS AND SUB–SYSTEMS

- Nuclear Reactor Fuel Handling System
- Automatic Mining Machines (Joy & FMC)
- Remote Oceanographic Teleoperators
- Medical Remote Diagnosis
- Computer Controlled Teleoperator
- General Manipulators
- Medical Orthotic & Prosthetic Manipulators

Remote Teleoperator Systems

Nuclear Reactor Teleoperators. An advanced teleoperator nuclear fuel handling system, reactor inspection system and reactor service system has been developed by Gulf General Atomic. The nuclear fuel handling system and manipulator is normally computer controlled. In the event the reactor core shifts, the computer commences a "search-and-find" routine until the initial core element is relocated. Subsequently, the computer updates the coordinates of all other elements with respect to the location of the initial element. A manipulator mounted on an articulated beam with TV viewing provides reactor core inspection system capable of covering nearly 98 % of the primary reactor systems. A service teleoperator system following an equipment installation/removal sequence was devised to remotely remove and replace a steam generator module within the reactor plenum.

Remote Automatic Mining Machine. In response to demands for increased productivity and safety, Joy Manufacturing introduced a completely automated coal mining machine in 1960. The mining machine, consisting of a boring machine, conveyer train and carousel conveyer storage mechanism, was successfully operated remotely from outside the mine shaft. Productivity was increased from approximately 30 tons per manshift to about 200 tons manshift.

Remote (Inherently Safe) Mining Systems and Procedures. The U.S. Bureau of Mines awarded a 5.5 million dollar contract to FMC Corporation in 1971 to develop continuous and conventional coal mining equipment and systems which will reduce accidents in underground bituminous mines. Phase I of this contract has concentrated on three important aspects of safer equipment: the development of standardized equipment controls and safety cabs for on-site operator protection, the investigation and development of remote control and viewing equipment, e.g., a teleoperator roof-bolting machine, and the investigation of remotely-operated continuous conveying equipment and procedures.

Remote Oceanographic Teleoperators. Scripps institution of Oceanography has developed and operated numerous undersea teleoperator systems. Scripps Marine Physical Laboratory designed the Benthic Laboratory for in situ ocean floor experiments and to support a current meter sensor bank on the Scripps Canyon floor at a depth of 460 meters. The RUMM vehicle equipped with manipulator, acoustical imaging, TV cameras and tactile sensing capability has been employed in ocean floor survey and remote coring operations at depths of 1500 meters. DEEP-TOW employs an ocean floor, teleoperator survey vehicle, which tows its surface support facility, while remotely surveying the ocean floor. The Scripps Deep Drilling Project has strongly impacted the science of geology offering new evidence of evolutionary mechanisms and mineral wealth beneath the oceans' and seas' floors. This project has developed and refined deep, remote drilling techniques and reentry capability at depths of 6000 meters. Scripps Visibility Laboratory has researched and improved both underseas and atmospheric sensory feedback so essential to teleoperator control.

Teleoperator Control/Communication Subsystems

Computer Aided Teleoperator System. The Department of Computing and Information Services, Case Western Reserve University, has developed a computer controlled manipulator system under NASA/AEC contract which is capable of remotely disassembling a nuclear reactor model. In addition, this system is capable of obstacle recognition, sensing obstacle space and subsequent obstacle avoidance, while selecting optimum path to work target.

Remote Medical Diagnostic System. While suggesting a program in medical data telecommunications, the Stanford University Medical Center EEG data transmission and analysis system work-in-progress and proposed work represents a necessary precursor to a total health care system employing the integration of telecommunication and teleoperation. Large segments of the world's population do not have access to professional medical diagnosis and treatment. Remote areas such as Alaska, India and many of the Pacific Islands could be medically linked to outstanding medical centers of the world by dedicated satellites initially providing transmission of medical data and two-way TV, and ultimately enabling remote teleoperator patient treatment. A modest demonstration project has been outlined by Stanford Medical Center applying state-of-the-art technology to the development of a remote medical capability.

Manipulator Devices

General Purpose Manipulators. Programmed and Remote Systems Corporation and Central Research Laboratories have designed manufactured and sold manual and powered manipulator arms, control systems and remote handling devices for many years. Their manipulators range in size and capacity from laboratory scale master-slave, mini-manipulators to massive powered manipulators capable of handling loads of 100 kilograms. These manipulators were designed to simulate man's arm, to allow man to remotely handle hazardous materials or to operate in a hazardous environment controlled from a safe, remote area. Initial application for manipulator devices was found in nuclear hot labs and subsequently broadened to include other "hostile" environments — for example, underseas and outer space applications such as the Surveyor moon sampling manipulator.

MB Associates has developed a manipulator arm which represents one of the most advanced and dexterous devices available today. It is a master-slave system; the slave arm being electronically controlled either by an exoskeleton master worn by its human operator, or by a small minicomputer. The system is the first-of-its type, man-equivalent, nine joint, position servo controlled, hybrid electro/mechanical/hydraulic/pneumatic teleoperator. It is a master-slave system, the slave being electronically controlled by an exoskeleton master, worn like a "coat" by its human operator located at a remote non-hostile control site. Vision of the slave work area is provided at the control site by an operator controlled, head-aimed stero TV system.

The NAT design incorporates force feedback which allows the operator to sense the weight of objects he handles with the slave, as well as control the gripping force of the slave. The system is sensitive enough to stack raw egges, to pour liquid from one fragile glass to another or to handle 10 kilogram, 18 centimeter diameter objects at arm's length. It can also be set up to amplify the force exerted by the operator to prevent fatigue when handling heavy loads over extended periods of time. Extremely delicate tasks such as threading a very small needle and assembling electronic circuit boards were performed for the first time by any teleoperator system.

Medical Manipulators. There are only a few institutions and medical centers in the country that specialize in the development and improvement of prosthetic (artificial limbs) and orthotic (powered extremity braces and actuators) devices. Rancho Los Amigos Hospital has perhaps led the field in advancing the state-of-the-art of orthotic rehabilitation arms. The present Rancho electric arm, which is externally powered, may be proportionally controlled by residual muscle-power, or, in the case of total paralysis, by multichannel tongue operated control switches. Of significance to broad-scale rehabilitation is the relatively low cost of these arms — about $ 2,000 each. In addition to direct medical application, Rancho has built numerous externally-powered teleoperator arms for NASA universities and industry.

NASA's Ames Research Center has devised a hard-suit for manned space application. This hard-suit has recently been internally-powered for potential application as a teleoperator/robot. Recent medical interest has suggested this hard-suit, if externally-powered, would have extensive application as an exoskeleton for a paraplegic patient. Multiple patient exercise therapy could also be possible using the NASA exoskeleton hard-suit.

The aforementioned teleoperator systems, subsystems and manipulators are presented as examples of present state-of-the-art technology. This limited survey serves only to provide a technological overview which should be contrasted with the three areas of user needs discussed earlier.

SUMMARY

This survey study indicated that Teleoperator/Robotics technology has not been vigorously pursued in the United States to date. Application of teleoperator devices has been restricted primarily to nuclear material and equipment handling techniques, deep-diving submersibles and limited prosthetic and orthotic rehabilitation. Outstanding work is being carried out currently by numerous groups throughout the country. Unfortunately, there is an almost total lack of coordination and integration of the work of these groups — coordination and integration so necessary to weld the many subsystems together into a total advanced system capable of meeting broad user needs and yielding the benefits promised. No industry, government agency or professional society has attempted this coordination and integration. The result is that most of the available hardware and controls are highly specialized, and do not lend themselves to broad diversified application. Herein are significant national benefits to be realized.

That almost one American out of ten is either physically handicapped or disabled in varying degree is a national liability. Not only are we losing the productive potential of these individuals, but in the case of the more seriously handicapped, additional able-bodied persons must be committed to their partial or full time care. Considering these factors: (a) our national empathy for the underprivileged and handicapped; (b) our need to remain productively competitive in the international marketplace; and (c) the existence of potential rehabilitation technology, it would appear that further development and application of teleoperator/robotic technology is critically important.

A review of our present and future national energy requirements provides another significant argument for the early development of a national teleoperator/robotic technology. At present we rely principally upon coal to supply our fossil fuel fired power plants. Because our national conscience demands that miners should not be exposed to the hazardous, often fatal, conditions inherent in deep mining operations, we are imposing increasingly severe health and safety standards upon this industry. The effect, however, in the absence of major technological break-throughs, has been reflected in significantly increased mining costs and coal prices and lowered productivity. Ultimately, teleoperated mining equipment should be developed which would remove the miner to a safe, remote location away from the work site, and which conservatively promises a tenfold increase in coal production per man. As more nuclear power plants come into existence, replacing present fossil fuel power plants, we must again place greater reliance upon teleoperator technology. Man cannot be exposed to the radiation environment that a machine can safely and easily tolerate. An entirely new generation of sophisticated teleoperators/robots with practical, computer-controlled artificial intelligence will be required to safely handle nuclear fuel at the reactor and in processing, and to inspect and perform routine and emergency maintenance and repair during the operation of these power reactors.

In view of our increasing awareness of the finite nature of our biosphere's resources, we also are going to be forced to explore and exploit the oceans and ocean floors in order to assure an adequate supply of essential minerals, hydrocarbons, minerals-rich soils and possibly food. Man has learned in the "60's how to function in the vacuum of outer space; in the '70's man will be required to either function under hundreds of atmospheres of water pressure, or to develop teleoperator/robots which can extend man's reach into the hostile environment of the deep oceans. As the present, only the latter holds significant promise.

For these reasons, this study recommended that the Federal Government undertake to establish and coordinate an integrated program in teleoperator/ robotic technology to demonstrate the productive advantages to be derived from the broad-scale systematic application of this technolgy.

SPECULATIVE TELEMATION

Based upon this survey of state-of-the-art telemation in the United States and its application to specific areas of user needs, I would now like to offer some speculations on impending developments in the extension and application of telemation to remote emergency medical care and to remote mining operations.

Use of Manipulators For Remote Medical Emergency Care

Recent advances in the technology of human extension by remote manipulation , concepts derived largely from the U.S. space program, make possible professional medical care, including surgery, in situ at the emergency site and remote from an environment specializing in acute care such as a hospital emergency facility or clinic. Figure 1 illustrates a remote surgical system.

Present emergency care even at accessible accident sites offers little more than first aid treatment by medically unskilled personnel (police, fire and ambulance). The injuried person is transported, usually to the nearest hospital, which may or may not have 24 hour emergency treatment capability. Unfortunately, many ambulances and their operators are not equipped, nor are they required, to provide lifesaving functions such as maintenance of respiratory exchange, blood flow, or skeletal stability. It is small wonder that about 20 million Americans are physically disabled today, and that there are in excess of 135,000 trauma shock deaths (55,000 D.O.A.'s) recorded annually. Yet this nation has a higher per capita ratio of skilled doctors than any other developed nation except Israel. The problem is to extend the doctor's reach and unique abilities to the patient without the time delay normally associated with emergency care, thereby substantially reducing shock trauma resulting from improper handling at the emergency site or in transit.

There are also many instances where the patient is so remote from medical facilities as to preclude medical care. Serious medical emergencies that occur aboard merchant ships at sea often require hours (by helicopter) or even days before the patient receives more than symptomatic treatment; transfer of such patients in rough seas cannot be effected without introducing severe shock trauma. Mining and similar disasters frequently place injured but alieve persons beyond the reach of conventional treatment for days. Natives of the under-developed remote regions of the world (even including certain areas of the United States and Canada) are frequently isolated from medical care.

A proposal, currently under funding considerations, suggests the use of human extension through remote manipulation to upgrade the quality of medical care at both remote and accessible areas, and to make the most effective use of scare skilled medical practitioners throughout this nation and the world for the relief of human suffering. This proposal is directed specifically to a one year pilot study of the use of human extension by remote manipulation for extended medical care at the emergency site. It includes a survey on site remote medical care capability. the design of an emergency care package and a pilot research demonstration of remote surgery for one or more procedures.

Initially assuming a modular concept. a proven minimum requirement for remote patient treatment is a two-way color TV/voice communication system (Massachusetts General Hospital's Logan Airport Emergency Room). which enables skilled doctor/patient interfacing. Next, in order of importance. basic diagnostic equipment is required at the emergency site: respirometer, sphygmomanometer, remote stethoscope, fluoroscope, image intensifier radiography, EKG machine and possibly EEG capability. All such equipment must be remotely operable through reliable telemetry, which state-of-the-art technology presently permits. The ultimate modular design would provide remote treatment capability, eliminating the need for trained paramedical personnel to be in attendance at the remote emergency site. Envisioned for this purpose would be a pair of master/slave manipulator arms remotely extending the doctor's manipulative, tactile (touch), visual (stereo TV or fiberoptic telemetry at terminal "hands") and listening (stereosonic) capability. Adjunct treatment equipment would provide for remote maintenance of life support functions and might include respirator, heart stimulator, blood flow control apparatus, anaesthesia equipment and skeletal support systems. In summary, an idealized modular remote care system must provide the attending physician with reliable two-way communication, diagnostic and treatment capability in that order. Each of the preceding concepts represents a distinct and measurable improvement over existing emergency care procedures.

As remote emergency care is demonstrated to be feasible, and the concept becomes accepted domestic medical practice, a number of international implications will become evident. As emerging nations place greater emphasis on developing technology — the greater use of machinery, cars, etc. — they will encounter an increasing accident frequency. A frequency they may be unable to deal with affectively unless they are able to concurrently educate and develop a medical staff adequate to their needs. A practical interim solution would be to draw upon medical staff in the developed nations to provide that care using remotely-operable facilities linked by dedicated medical communication satellites.

Further, these remotely-operable emergency care facilities when not in direct use could be used for medical education purposes, linking the outstanding medical centers of the world to the most remote regions, providing access to the most advanced medical techniques and procedures available.

Remote Mining by Telemation

The United States Bureau of Mines began soliciting proposals early this year for innovative systems concepts related to the underground mining of coal. These concepts were to be aimed at improving the health and safety of the miner, and at improving productivity. At the time this paper was submitted, the following proposal had been discussed with the Bureau for possible funding.

An incremental approach towards the achievement of a totally remote, teleoperated mining machine was suggested. Initially, current continuous mining machines would be equipped with remote sensors, operating controls and stereo visualization capability (one way TV). Such equipment would be controlled by its human operator from a remote, relatively safe station far removed from the active mining face.

Experience with this remotely operable mining equipment might lead to the design and development of a totally remote miner, described in this proposal as a logical extrapolation of present mining technology consistent with both societal, environmental and productivity requirements.

The totally remote, underground mining machine and conveyor train will be remotely controlled from a block-house control center above ground. For optimum remote guidance and control, the drilling, explosive loading, ore-loading and initial conveying operations should be consolidated on a single, caterpillar-crawler, diesel or electric powered vehicle similar to that shown in Figures 2 and 3. The ganged-drills advance into the rock working face through holes in the indexed blast shield/loader. Following drill retraction, the shield indexes the explosive loading tubes. The explosive, similar to 3/4 inch diameter primacord, is fed into the loading tubes from continuous reels in appropriate lengths. The nose of each section of primacord is capped with a pointed steel conical head with spring-steel flutes which dig into the drill hole and hold the length of primacord in position as the loading tube retracts. The rear end of the primacord is capped with a detonator and sleeve which is armed as it uncouples from the nose of the next length of primacord. After drilling and loading the explosive at the working face, the mining rig backs off, the charge is detonated by radio wave, and the machine returns to commence ore loading. Mechanical or hydraulic jacks raise the front end of the bucket/hopper/conveying system and the bucket loads the broken ore onto the loading conveyor. This conveyor serves also as a protective canopy against possible roof falls. In which event, the conveyor has sufficient strength and capacity to remove the fallen rock so that the mining machine may be backed out of the critical area. It should be noted this machine could also support an automatic roof-bolter if necessary to avoid roof falls. The mining machine conveyor finally discharges into a second hopper which discharges onto the narrower continuous conveyor train. The latter is advanced by the mining machine and individual drive motors. All drive motors are variable speed, oil-cooled with sealed carbon alloy bearnings. Drills provide force and heat-temperature feedback to the remote control station. Remote broad-spectrum sensors along with stereo TV cameras provide total visualization at the working face. Drill-angle and subsequent ore-loading determine the horizontal or inclined path the machine is directed to follow. Most of the ore-tracking and machine guidance is remotely and automatically controlled by a central computer located in the blockhouse. A manual executive override mode is available to the operator for emergency situations.

Conventional mining methods can be expected to evolve along lines already indicated. With increased emphasis on worker safety and comfort, greater productivity and environmental protection, it should be assumed that most mining operations by the end of this decade will be remotely controlled, highly-automated and characterized by at least an order of magnitude greater productivity and scale of operation.

It should be recognized that the comprehensive application of telemation to the fields of medical care and mining will undoubtedly create systems far superior to those mentioned here. Further speculation is unnecessary to illustrate the manner in which properly applied telemation may improve the quality of life and the national economy.

Table I – ESTIMATE OF DISABLED PERSONS IN THE UNITED STATES (1971) (*)

DISEASE CATEGORY	TOTAL NUMBER OF PATIENTS	TOTAL NO. IN REHABILITATION CENTERS	NUMBER OF NEW PATIENTS/YEAR	PERCENT REHABILITABLE	NUMBER THAT COULD BE HELPED WITH CURRENT TECHNOLOGY
Stroke	2,000,000	250,000	500,000	60	400,000
Cerebral Palsy (**)	550,000	?	50,000	45	500,000
Multiple Sclerosis	500,000	?	50,000	?	50,000
Spinal Injuries*	100,000	1,000	10,000	95 (80 to jobs)	100,000
Amputees	350,000	?	?	100	350,000
Diabetes	3,000,000	?	?	?	600,000
Rheumatoid Arthritis	13,000,000	?	?	?	1,000,000
TOTALS	19,500,000	–	–	–	3,000,000

* Liberty Mutual estimates total cost for each quadruplegic patient, ranges from $ 250,000 to $ 350,000; direct medical treatment costs range from $ 25,000 to $ 35,000 per patient, the remaining costs cover such things as workman's compensation, extended care, etc.

(*) Estimates supplied by Dr. James Reswick, Rancho Los Amigos Hospital, assembled rom sources ranging from "hard" to "soft". Orders of magnitude appear correct.

(**) 1956 Figures

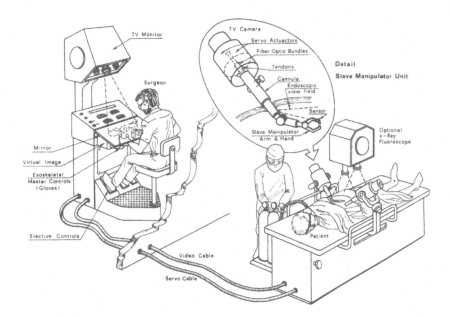

Fig. 1   Schematic representation of speculative remote surgery

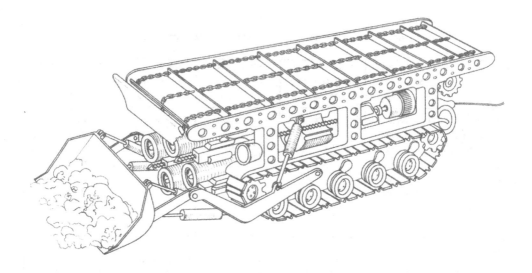

Fig. 2   Speculative remote mining machine

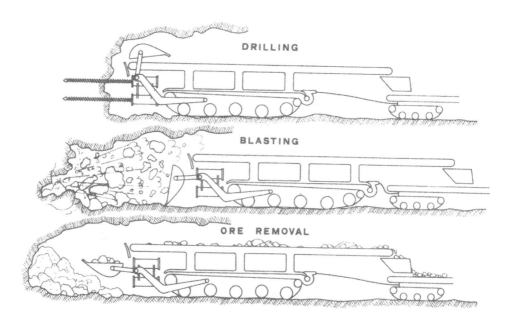

Fig. 3   Speculative remote mining operations

THE APPLICATIONS OF THE REMOTE CONTROL OF THE MANIPULATOR IN MANNED SPACE EXPLORATION

Dr. Stanley DEUTSCH
National Aeronautics and Space Administration
Washington, D.C. 20546

Dr. Thomas B. MALONE
Essex Corporation
303 Cameron Street
Alexandria, Virginia 22314

(*)

(*) All figures quoted in the text are at the end of the lecture

BACKGROUND

The concept of remote control for space applications has been receiving increasing emphasis over the past several years, both in the United States and in the Soviet Union. Russian efforts in this area have produced the first (and second) remotely controlled lunar rover, the Lunakhod system. This paper will discuss systems and technology development efforts included in the United States Aerospace Remote Control Program.

In order to differentiate systems which are under some degree of manual remote control from the more automated robot systems, the former have been designated by the National Aeronautics and Space Administration (NASA) as teleoperator systems. A teleoperator system has been officially defined, by NASA, as a remotely controlled, cybernetic, man-machine system designed to augment and extend man's sensory, manipulative, and cognitive capabilities. The NASA does not propose the use of teleoperator systems to replace man in space. Actually, the intention in the development of teleoperator systems is to provide an additional alternative to the approach of placing the man physically at the worksite. In situations where this approach compromises the safety of the man, or requires capabilities which are beyond those possessed by the human operator, or where man's physical presence is not yet feasible, the use of teleoperator systems becomes a viable alternative to locating the man at the site.

The essential subsystems of a teleoperator system include the man, the remotely controlled device located at the worksite, and the communications link between these two (Figure 1). The man is always in the control loop of a teleoperator system, as an active controller or as a supervisor, managing the performance of the system under some form of computer control. The remotely controlled device is the machine which actually performs the work for which the system is designed. This device incorporates an array of manipulative and locomotive systems, sensors and feedback systems and guidance systems. The communications link provides for transfer of commands and control inputs from the man to the machine, and for relaying sensed information and feedback data back from the machine to the man.

The teleoperator system thus incorporates many of the advantages of manned systems on the one hand, and mechanical systems on the other. Since man is always in the control loop of a teleoperator system, the system is provided the decision making and adaptive intelligence capabilities which are uniquely human. Conversely, since the actual work is performed by the remotely controlled device, the system incorporates the durability, strength, and the expendable nature of a machine. Use of a teleoperator system is generally safer than placing the man at the worksite, and is generally more flexible and adaptable than an automated mechanical system.

While the designator "Teleoperator" is of recent vintage, the use of teleoperators is not a new development. The Atomic Energy Commission has used manipulative systems extensively for the remote handling of radioactive materials. The Department of Defense has been expending a good deal of research and development in areas of remotely piloted vehicles and remotely underwater systems. Even within NASA, teleoperator systems have been developed, such as the stationary lunar surface sample, the Surveyor, and the film retrieval support system for Skylab, an extendable manipulator.

While such systems have existed in the past, those systems were usually designed for a special purpose. The emphasis was placed on development of operational teleoperator systems rather than on the development of teleoperator technology. Today we are witnessing a concerted effort to systematically develop, integrate, and advance the technologies required for teleoperator systems over a wide range of space applications. This growing concern for technology development will result not only in more effective systems but also in more economical systems. The costs of system development will be reduced by providing the system design engineers with a well established technology base to draw on.

The remaining sections of this paper will discuss potential earth orbital space missions for teleoperator systems, candidate aerospace teleoperator systems to perform these missions, and the current status of teleoperator technology development within NASA.

Teleoperator Space Missions

The use of teleoperator systems in the United States space effort is presently being investigated for earth orbital science and application programs, and for planetary exploration missions. This paper will be concerned primarily with the earth orbital systems. The space shuttle program will provide the first opportunity to fully exploit the teleoperator system as a means of performing and supporting operations in earth orbit. Teleoperators in earth orbit are generally considered to be support systems, supporting some facet of the mission of a primary system.

For the earth orbital teleoperators, there are two general classes of missions: spacecraft support missions, and experiment support missions. Spacecraft support missions refer to the situations where the teleoperator is required to enable or enhance the performance capability of a spacecraft. The systems to be supported include the shuttle itself, and shuttle payloads (satellites, sortie labs, and station modules). Shuttle support missions under consideration for teleoperator systems include heat shield inspection, contamination measurement in the vicinity of the orbiter, shuttle servicing (refurbishment, repair, and on-orbit maintenance) and astronaut support. Teleoperator payload support missions include payload capture and retrieval to the orbiter, payload deployment, and payload inspection, assembly, and servicing. Experiment support missions being investigated for teleoperators include activation, operation, and monitoring of experiments, and acquisition of experimental data. Projected missions for earth orbital teleoperators are summarized in Table 1.

Requirements associated with shuttle support missions are currently being established. Little is known today concerning these requirements beyond the fact that some inspection, contamination measurement, and orbiter servicing will probably be required.

Requirements related to payload and experiment support missions are more readily available. The NASA 1972 shuttle mission model identifies a total of 288 missions for 68 payloads in the 1979-1990 timeframe. Characteristics of these payloads are presented in Table 2.

As indicated in Table 2, a considerable majority of payloads identified in the NASA mission model will require support in terms of retrieval, servicing , deployment, experiment support, etc. The capability to provide such support actually is one of the primary justifications for the shuttle, that it will enable low cost space operations through recovery, refurbishment, and reuse of payloads and will itself provide the means for placing payloads into orbit.

Teleoperator missions under consideration for support of planetary explorations include exploration itself, and experiment support. Exploration missions include remotely controlled maneuvering over the surface of a planet, and sampling of the surface, subsurface, and the environment (atmosphere, illumination, etc.) of the planet. Experiment support missions include emplacement of experiments, activation and operation of the experimental packages, and servicing, inspection, and checkout.

Earth Orbital Teleoperator Systems

The systems under consideration for earth orbital teleoperator missions include two general classes: free flying systems, and attached systems. The free flying systems include the Free Flying Teleoperator System (FFTS), and the Space Tug. The attached class of systems comprises the Shuttle Attached Manipulator System (SAMS) and space station attached teleoperators.

1. Free Flying Teleoperator System

The Free Flying Teleoperator System represents a relatively small unmanned space vehicle controlled by an operator in the shuttle. The system is conceived of as containing propulsion systems, stabilization systems, video sensors, manipulator systems, and communication devices (Figure 2). The FFTS is designed primarily for operations in the vicinity of the shuttle, however, it can perform payload support missions in geosynchronous orbit when provided with the additional propulsive capability of the Space Tug.

A conceptual FFTS has been developed by Bell Aerospace for NASA Marshall Space Flight Center. This system weighs 400 lbs. and is approximately three cubic feet in size. The propulsive impulse of the baseline system is 15,000 lb.-seconds. The system is limited to a maximum range of 10,000 feet from the orbiter and is capable of retrieving payloads of up to 7,000 lbs. from distances up to 2,000 feet from the shuttle. The working end of the FFTS is of modular design and an array of capture grapplers and servicing manipulator concepts are being investigated. Stabilization and orientation of the baseline vehicle is provided by three rate gyro assemblies and three control moment gyros. Guidance is provided by ranging sensors and by video sensors. The control station of the FFTS may initially be located in the sortie lab on the shuttle, for FFTS alone missions, and on the ground for FFTS-Tug missions in orbits which the shuttle cannot attain (high and geosynchronous).

The primary applications of the FFTS for shuttle and payload support missions are in the areas of retrieval, servicing, experiments support, and shuttle support . A recently completed program to analyze FFTS mission applications, conducted jointly by Bell Aerospace and the Essex Corporation, indicated that the FFTS is applicable for retrieval of 31 of the 40 payloads requiring or desiring retrieval (77 %). This same study reported that the FFTS is applicable for servicing of 51 of the 67 payloads (88 %). Servicing missions for the FFTS include systems update, removal/replacement of failed modules, and array deployment or retraction.

In the experiment support area, the FFTS is a candidate system to serve as the subsatellite associated with seven shuttle payloads. For shuttle support missions, the use of the FFTS ensures visual access and reach to inspect and service the entire external surface of the shuttle. The FFTS also provides the capability for taking contamination and plasma wake measurements at ranges in excess of 50 feet from the orbiter (i.e., beyond the reach of the shuttle attached manipulators). Finally, the FFTS has application in rescuing an EVA astronaut (if required).

A significant capability afforded by FFTS in the capture and stabilization of dynamically unstable spacecraft without danger to the shuttle crew. Various methods for capturing unstable satellites using the FFTS are under study.

A second advantage of the FFTS is its extended range. As indicated above, the FFTS alone can operate at ranges up to 10,000 feet from the shuttle. When configured with the tug, its range is extended to include geosynchronous altitudes.

The FFTS also offers the shuttle an additional magnitude of flexibility of operations in that the orbiter vehicle position and orientation is in no way constrained. The FFTS also adds to the shuttle a new dimension of efficiency since the orbiter is free to perform other missions on orbit while the free flyer performs its mission.

The significant limitations of the FFTS are seen in its inability to emplace a payload directly into the shuttle bay, and in its mass handling capabilities. The FFTS must rely on a recovery system such as the attached manipulator or an automated recovery mechanism to emplace a retrieved payload and to recover the FFTS itself to the bay. The mass handling limitation results from the propulsion requirements associated with transfer of large payloads. The FFTS is capable of translating a large mass, such as the large space telescope, only if additional propulsive capability is provided. No such limitation exists for the shuttle attached manipulator, which can handle payloads of any mass up to 65,000 pounds.

2. Shuttle Attached Manipulator System

The shuttle attached manipulator system (SAMS) represents one approach for the shuttle cargo handling system (Figure 3). One concept under study includes two 50 foot articulated booms, and effectors, visual and possibly tactile sensors, and an operator station located immediately forward of the cargo bay. Each boom is mounted at the forward end of the cargo bay, and consists of three segments 15 inches in diameter. The operator station is provided with controllers for control of the booms and end effectors, with windows for direct viewing of the bay and the manipulator working area, perhaps augmented by video displays for work away from the direct viewing areas.

The SAMS may be used for the deployment of payloads from the cargo bay and emplacement of payloads into the bay. The booms can be designed to handle payloads weighing up to 65,000 pounds. In the preliminary description of the first 21 shuttle missions prepared by Marshall Space Flight Center (MSFC), the SAMS is identified as supporting the retrieval or deployment activities in seven missions.

In addition to the basis deployment and retrieval missions, the SAMS also has application for payload servicing, either while the payload is docked to the orbiter's docking adapter or while it is held in place by one boom while being serviced with the other. In the first 21 shuttle missions, the SAMS may perform a payload servicing mission, specifically, inspection and module exchange for the Long Duration Exposure Facility on flight six.

The SAMS has application for experiment support since it is capable of such operations as deploying contamination sensors or plasma sampling systems and manipulating these devices as required. The system is also a feasible candidate for direct shuttle support missions since, through its sensors (primarily visual), it can perform inspection, damage assessment, and verification of systems readiness for re-entry, it can be used to assist rescue of an EVA astronaut as well. Some consideration is currently being given to the use of the SAMS to support EVA and even to provide the mobility system for the astronaut.

The SAMS provides the shuttle with a general purpose, highly versatile and adaptable payload handling system. It can handle payloads over a wide range of sizes and shapes provided that an adequate interface has been provided. Normally, only one arm would be employed at a time. The SAMS can be designed so that it can be quickly disconnected and jettisoned from the shuttle in the event of failure or the occurrence of hazardous dynamic conditions of the payload being handled. The baseline system requires one operator and techniques of computer control, primarily of the cargo bay emplacement operations, are currently being investigated.

The primary limitations of the SAMS are its reach capability and its constrained capability to capture dynamically unstable free flying payloads. The reach limitation refers to the fact that the only payloads which are accessible to the SAMS are those located within 50 feet from the shuttle at the time of retrieval. Since the shuttle cannot attain some orbits (high or geosynchronous), payloads in these orbits cannot be retrieved directly using the SAMS. The limited capability of capturing dynamically unstable payloads results from the response rate of the system . The SAMS is incapable of capturing a large tumbling payload if the rate is one RPM or greater. The SAMS will be capable of capturing and stabilizing payloads in a pure spin , but this capability degrades sharply if nutation as well as spin is present in the payload dynamics. The capability of the SAMS to inspect the shuttle is also limited by the reach envelope of each boom. As mounted at the fore end of the bay, the manipulator tip can provide visual access and manipulative access to almost the entire upper surface of the orbiter, but is limited in viewing or reaching the tail and underbelly.

Another potential limitation of the SAMS is that it requires handling of payloads in close proximity to the shuttle. If the payloads are contamination sensitive, and if the space environment immediately adjacent to the shuttle is composed of particles and gases emitted from the orbiter, handling of the payloads in and through this environment poses a hazard to the experiments. This restricts the performance of these experiments near the shuttle.

Earth Orbital Teleoperator Technology Development

The status of development programs for the FFTS and the SAMS is described in the following sections.

Technology for the FFTS and Tug systems is under development at Marshall Space Flight Center. Technology areas of primary concern include the manipulator-grappler system, the control system, the video system, and the mobility system. Manipulator and end effector, and control system concepts are currently being developed and evaluated through an intensive development program at Marshall Space Flight Center. Video system requirements and concepts are being established at MSFC by Martin Marietta for MSFC. Mobility systems and requirements are being developed and evaluated by MSFC on an inhouse basis. FFTS man-machine interface requirements are being developed by the Essex Corporation, for MSFC.

The major technology problems for the FFTS include manipulator-grappler controllers and control systems and feedback systems. Controller concepts range from master-slave anthropomorphic controllers, which reflect the configuration of the manipulator arms themselves, to computer aided control of manipulator tip position and orientation in space. Controllers must be capable of providing control of the position, rate, orientation, and configuration of the manipulator arm and of the end effector.

Feedback systems include visual and kinesthetic. Visual systems include cameras, displays, lighting, and display aids at the monitor and at the target. Kinesthetic feedback includes display of applied forces, manipulator configuration, and contact faces.

The primary technology development areas for the SAMS are in the areas of stabilization, structure, and manual control. The design criteria for the manipulator system as reported by the NASA Johnson Space Center, Houston, Texas, indicate a two inch tip placement accuracy for the system. The requirements for providing this accuracy with a long boom system having an inherent long period oscillation represents the major technology problem for the operator control of the SAMS. Potential solutions to the oscillation problem lie in increasing boom rigidity and in providing an active damping or counter-force system.

A conceptual SAMS has been developed by Martin Marietta for JSC. This system incorporates the capabilities described above and includes advanced concepts for manipulator control. The control of an articulated manipulator has been a problem for teleoperator systems even on earth. The control problem is usually resolved through development of an anthropomorphic controller which reflects the configuration of the manipulator in a master-slave arrangement. A second solution gaining acceptance among teleoperator design personnel is the use of the computer to configure the manipulator segments to place the tip at a point in space as commanded by the controller. The early Martin approach was the master-slave arrangement. A later proposed concept uses the computer to control the boom while the man controls the tip.

Technology development for the SAMS is being conducted at NASA JSC. Technology for earth orbital teleoperator systems in general, including primarily visual systems, feedback systems, and end effectors, is being developed at MSFC. Technology for video systems and tactile sensors is being developed at the NASA Ames Research Center.

Space Teleoperator Technology Applications

As is apparent from the descriptions of earth orbital teleoperator missions, systems, and technology development programs, the concept of using remotely controlled systems in space is gaining acceptance throughout NASA. As technologies are being advanced to enable development of earth orbital teleoperator systems, consideration is being given to the ways in which this technology base has applications for solving problems on earth.

The NASA Office of Technology Utilization is already deeply involved in identifying and implementing teleoperator technology applications to the medical engineering profession. Concepts for using space manipulator technology for advanced prosthetic and orthotic systems are being established. Applications of remote control technology for control of the environment and of aids for the severely handicapped are being identified. Teleoperator sensor technology, notably visual systems and tactile systems, is being applied to resolve problems of the sensory deficient. The broad technology base associated with the remote conduct of space operations is being analyzed for potential application to the remote control and monitoring of health care delivery services.

Remote control systems technology being developed for space flight is also being applied to the resolution of pressing problems in such diverse areas as coal mining, undersea operations, fire fighting, and manufacturing and production processes. The technologies developed for manipulator systems, mobility systems, remote control systems, and sensor systems, for earth orbital operations, will receive increased attention for application to other problems as these technologies become integrated, validated, and documented.

TABLE 1

Candidate Earth Orbital Teleoperator Missions

Shuttle Support Missions:
    Inspection (heat shield, payload bay, etc.)
    Damage Assessment
    Contamination Measurement
    Shuttle Servicing (refurbishment, closing of doors prior to re-entry, etc.)
    Astronaut Support (rescue, operational assistance)

Payload Support Missions:
    Retrieval (including capture, transferç recovery)
    Retrieval Support (including safing and securing)
    Deployment (retraction form bay, transfer, emplacement in orbit)
    Deployment support (spin up, checkout, preparation, shrouding)
    Servicing (maintenance, repair, update, refurbishment)
    Assembly (module mating, erection)
    Inspection (survey, surveillance, visual check)

Experiment Support Missions:
    Experiment checkout, deployment, assembly, setup
    Experiment activation, operation, and monitoring
    Data acquisition and measurement

TABLE 2

Characteristics of Payloads
in the NASA Mission Model

Characteristic	Number of Payloads	Number of Missions
In Low Orbit	34	188
Satellites – Observatories	17	84
Sortie Labs	8	56
Station Modules	9	48
In High and Synchronous Orbit	15	63
In Escape or Planetary Orbit	19	37
Requiring Retrieval	40	204
Requiring Servicing	63	267
Requirement Deployment	65	251
Requiring Experiment Support	19	107
Requiring Inspection	63	267

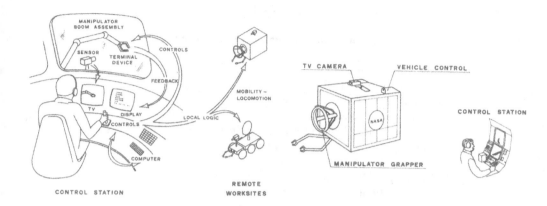

Fig. 1 Teleoperator system: a remotely controlled cybernetic man-machine system
designed to augment and extend man's
sensory, manipulative and cognitive capabilities.

Fig. 2 Free flying teleoperator

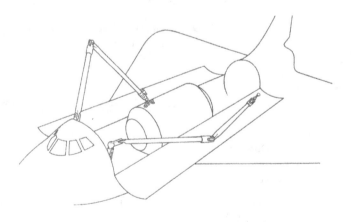

Fig.   3 Shuttle attached manipulator system concept

REMOTELY MANNED SYSTEMS FOR OPERATION AND EXPLORATION IN SPACE

Ewald HEER

Jet Propulsion Laboratory

Pasadena, California 91103

(*)

Summary

A brief overview is presented of Remotely Manned Systems with emphasis on their use as tools for exploration and operation in space. Remotely Manned Systems missions and functions in space are described and classified in relation to other existing or planned space systems. Problem areas of large-scale man-machine systems are identified based on experience in the Surveyor program, the Mariner 9 Mars orbiter project and the Apollo program. The effects of communication time delay on system performance are investigated using the average velocity of a Martian rover as performance indicator. A substantial performance increase can be achieved by providing certain autonomous capabilities to the remote system.

(*) All figures quoted in the text are at the end of the lecture

I. INTRODUCTION

During the decade of the 60's, man demonstrated the technological and organizational capabilities to explore space successfully. He expanded his horizons far beyond his natural earth environment mainly in two ways. Either he carried the environment required for his survival with him in devices such as space capsules or space suits, or he designed and used machine systems that augment and extend his sensory, manipulative and intellectual capabilities to these remote places enabling him to remain in a safe comfortable environment. Such man-machine systems consist of a remote system connected by command and feedback communication links to a manned control center. These systems, the subject of this paper, have been termed "Remotely Manned Systems" or "RMS" (Ref.1).

The RMS concept is more general than that of a "teleoperator". Teleoperators, a subclass of RMS, are general purpose, dexterous, cybernetic, man-machine systems, with one or more manipulators usually controlled and operated by only one human (Ref. 2). The terms "remote manipulator systems" and "teleoperator" can be used synonymously in many cases. The terms "robot" and "robot system" are often used for the remote system of an RMS inasmuch as it has autonomous motion or handling capabilities. A Remotely Manned System may have many humans in the control center and any required number of manipulators and/or robots at one or several remote places simultaneously.

This paper gives first a brief overview of Remotely Manned Systems concentrating on their use of tools for exploration and operation in space. Then, a description is given of various problem areas.

The first area is that of the man-machine system where the remote system has no automatic capability at all and all decisions for the total system are made by one or a group of humans in the control station on earth, and the second area addresses the questions of required automaticity in space based on operational efficiency, where the measure of efficiency is the average velocity traveled or the number of tasks accomplished during a certain time interval.

II RMS MISSIONS AND FUNCTIONS

The spectrum of possible applications of RMS or teleoperator systems in space ranges from various research and operational missions in earth orbit in conjunction with the space shuttle, space stations and satellites to primarily exploratory missions to the moon, planets, asteroids, comets, planetary satellites, etc. (ref. 3). Figures 1-4 give a schematic overview of some RMS systems.

The distinction between an operational RMS and an exploration RMS is not so much technical as one of intended function and can generally be characterized by four classes of RMS as shown in Table 1. The first one operates mainly on artifacts such as satellites, space shuttle and space station modules, orbital telescopes, and the like. The manipulators therefore handle objects which are (or should have been) designed appropriately for manipulation by the manipulators terminal devices. The interfaces between the terminal devices and the artifacts to be manipulated are thus controllable by the designer.

Operational RMS — The broad functions of an operational RMS are servicing, construction and emergency operations. Servicing operations include calibration, check out, data retrieval, re-supply, maintenance, repair, orbit change, replacement of experiments, cargo and crew transfer and recovery of spacecraft. Large telescopes, communication and experimental satellites, manned space stations, orbiting fuel depots, nuclear systems and the shuttle itself are obvious beneficiaries of such servicing functions. Construction in space includes the possible fabrication, assembly, deployment, emplacement and operation of large telescopes and particle counters, large communication systems, solar power stations, specialized manufacturing facilities, space station complexes, space exploration vehicles and the like. Special devices for rescue operations could be provided for space stations, shuttle and space tug. Such devices could also be on standby alert on the ground. The delivery systems need not be man-rated. They could deliver expendable life support systems or encapsulate the astronaut in a life support environment for return to a shuttle, space station or to earth. They could also perform first aid functions.

Exploration RMS — The broad functions of an exploration RMS are scientific data acquisition and information return to earth. The data acquisition includes such activities as environment sensing, sample identification, sample acquisition, sample analysis, experiment emplacement, operation and retrieval, and data storage. Many of the most important functions of an exploration RMS are therefore to deal with unknown environments and objects that are outside the control or knowledge of the designer. This is one reason why some exploration RMS should be designed so that the sensory subsystems, the manipulators and terminal devices at the remote place can respond adaptively to unknown, unforseen, and perhaps changing circumstances. Another reason is that future exploration RMS for space are expected to perform their mission functions far removed from earth where the telecommunication time delay plays an important role in their operation.

Flyby an orbital spacecraft are usually equipped with less general purpose capabilities than the other RMS classes. Their manipulative functions are rudimentary consisting of scanned platforms and the like. Nevertheless, recent developments show that in the future they will tend to have increasingly more general-purpose applications. A recent example is the Mariner 9 orbiter mission. Two spacecraft designed for television coverage of the Martian surface were launched; one launch failed. When the surviving spacecraft arrived at and went into orbit around Mars, a dust storm covered the Martian surface, almost entirely preventing visual access for several weeks. As a result, the previously established mode of exploration was changed adaptively, leading also to the photographic coverage of the Martian satellites Phobos and Deimos. The required adaptive maneuvers involved ground operations, the Deep Space Network and the remote spacecraft as an interactive "cybernetic" man-machine system. The general purpose aspect of the system was demonstrated, although in a limited fashion because of the constraints.

While most paths of motion for flyby and orbital spacecraft can be computed quite accurately based on a few determinate parameters before their missions, the path of motion for surface vehicles is dependent on many more parameters. These are usually not only unknown beforehand, but most of them can be determined quantitatively only after the vehicle is at the exploration site and after its sensors interact with the surrounding environment. The immediate interpretation of the sensor data may save the vehicle from disaster. For instance, knowledge of the bearing strength of soil in relation to the vehicle's weight may prevent it from sliding into a crater. Experience in this regard was obtained in the USSR lunar program when driving Lunokhod on the lunar surface. Although Lunokhod did not have manipulators on board, remote manipulators were used for the acquisition of lunar soil samples on the sample return missions of Luna 16 and Luna 20.

The only U.S. manipulators that have operated in space to this date were attached to the Surveyor landers, Surveyor III and VII, which were stationary platforms on the lunar surface. Considerable experience, including emergency operations in an adaptive mode to deal with a malfunction in another subsystem, was obtained in the remote operation of the Surveyor manipulators. Another remotely operated manipulator will be on the Viking lander on Mars in 1976 (Ref. 4).

III. GENERAL RMS CHARACTERISTICS

From the above discussion it is evident that certain commonalities exist among the various RMS classes, and that they can be discussed in terms of a total system model. The system depicted in Fig. 5 identifies the essential component systems required in a general complex RMS with many possible simplified variations. Examples of such complexity are certain free-flying remote systems and mobile surface explorers with on-board manipulators and end effectors, and with a propulsion or locomotion subsystem that enables these to move from point to point at the will of the operator in the control center or in response to autonomy incorporated in the on-board computer.

The sensor subsystem of the exploration RMS consists of various science sensors for numerous scientific investigations, and the engineering, navigation and pathfinding sensors that detect the information necessary to provide for the vehicle's safety and reliability. Clearly, some sensors are used for more than one purpose; for example, the video system is used for both navigation and scientific analyses. Similar statements hold for the complex operational RMS with the exception that in the design stage attention can be given to the identification and recognition of artifacts by suitably marking them.

The sensory information is processed by the on-board computer. Some of these data are required by the human operator to control and operate the system and are sent to the operator via the downlink communication transmitter, receiver and display. Before reaching the operator these data may be interpreted in combination with other pertinent information in the control center computer facility for suitable display. Other sensory information may be processed by the on-board computer to generate decisions for autonomous action by the remote system, if the on-board computer is equipped with at least some autonomous capability.

In the complex systems required for lunar and planetary surface missions, the "human operator" in Fig. 5 is represented by a complex organization structure in the control center. For instance, Fig. 6 shows the organization of mission operations for the Mariner 9 Mars orbiter. This mission provided the first opportunity to acquire and analyze information regarding the planet Mars while the spacecraft continued to obtain data, and to plan and execute mission operations in an adaptive manner over a prolonged time period. During peak mission activities, the number of persons involved and contributing within the organizational structure of Fig. 6 was more than 200. This is indicative of the managerial and operational complexity associated with such large-scale man-machine systems.

IV. EXPERIENCE WITH LARGE–SCALE MAN–MACHINE OPERATIONS

Three past projects were selected for an in-house examination: Suveryor, Mariner 9 Mars Orbiter, and Apollo.

Surveyor – The Surveyor program had the greatest amount of man-machine interaction of unmanned projects to date. It was remotely controlled, and involved complex operations over an extended period of time. The Surveyor spacecraft carried no on-board sequencer, computer, or stored commands for science operations.

Individual ground commands had to be sent for each operation. Commands could be sent at 0.5 sec intervals.

Mariner 9 Mars Orbiter – The Mariner 9 project involved the most complex and extended planetary operations yet carried out and is the first planetary mission to require much real-time interaction between scientists and operations. Mariner 9 carried a flight computer, and most ground commands transmitted information to be loaded into this computer for later use. Some commands were sent "directly" for immediate excution, and did not pass through this computer.

Apollo – The Apollo program has been the largest and the most complex overall. A first reaction may be that Apollo is not representative of an RMS mission. But from the point of view of the people in the ground control center, the equipment , operations and measurements they were concerned with were remote from them on the moon. That some of the operations on the moon were executed by astronauts did not basically change the RMS concept but puts it only into a category where the remote system is equipped with human intelligence as well.

Some problems arose in the interface between scientists and the spacecraft and ground hardware and software systems during operations. Others concerned the interface between scientists and operations people. Moreover, a number of problems which probably were basically in the areas of management, design, and reliability and testing occurred during science operations.

The more important problems brought out in the various areas included (1) flight hardware: science telemetry allocations; command structures; (2) ground hardware: computer capability; equipment reliability; return of overseas data by mail; (3) software: inflexibility and clumsiness; unreliability of operation and of data output; times required for command generation, forwarding, and sending; times required for data output; inadequate checkout; incompatibility of hardware with software; (4) procedures: inadequate testing, training, and simulation for science operations; lack of science operations contingency plans; information exchange between operations personnel and scientists; (5) humans: scientist attitudes; operations staff attitudes; scientists in unaccustomed roles; staffing schedules; last-minute hiring; human relations.

Despite these problems, all three projects were highly successful. The problems did, however, reduce the scientific output and are likely to be more serious if encounterd during operations of more complex future missions.

V. EFFECTS OF COMMUNICATION DELAY

The complexity and difficulty of operating an RMS is closely related to the communication distance and the resulting signal time delay between the control center and the remote system. Several distinct levels of operational difficulty may be associated with orders of magnitude of this distance, as depicted in summary form in Table 2. The complexity of the operational problem tends to expand hierarchically with distance. This complexity increases in discrete steps through the introduction of new problems at greater distances while problems at lesser distances carry over and frequently become more critical and more severe.

Aside from the signal transit time delay, a series of other time lags is developed in the communications feed-back and command loop. These include in the general case the times for: (1) sensor orientation, (2) reading sensor data into storage, (3) on-board data processing, (4) antenna orientation, (5) reading sensor data into the communications system, (6) ground-based data processing and control decision generation, (7) read-out of commands into up-link, (8) orienting vehicle and/or manipulators into the desired position, and (9) performing desired motions. The relative importance of these time intervals is strongly dependent on the kind of mission, the type of equipment, the mission operational organization, and the rate of data processing capacity of the various links in the control systems loop.

For an RMS without automatic capabilities at the remote system and at small distances when the communication transit time is not more than a tenth of a second, continuous closed-loop manual control is usually possible. However, at greater distance, the human operator is forced to abandon continuous control and go over to the move-and-wait open loop strategy, Fig.7. Operators initially find this situation upsetting but usually adapt quickly. In the move-and-wait mode, stability is then obtained by the inefficient use of time.

For operations from earth up to synchronous orbit, the total time delay is approximately 0.9 sec, which includes 0.2 sec for the time required for a single operator to respond, corresponding to the time interval (6) above, while two-way transit plus equipment time delay is included with approximately 0.35 sec. This distance belongs to the "critical region as shown in Fig. 7. Comparison of Table 2 with Fig. 7 gives a rough indication regarding those RMS that fall within the stable closed loop, the critical, or the stable iterated open loop or move-and-wait region. Only the Space Station attached, the Space Shuttle attached and the Space Tug remote manipulator systems would thus, with certainty, fall within the stable closed loop region. The inefficient expenditure of time using the move-and-wait strategy would apply to all RMS with distances greater than about 20,000 km between the control center and the remote system. The work at the remote place will then be accelerated and overall performance increased only by increasing the number of tasks or operations which the operator can command without feedback. Such an increase in performance can be accomplished by giving the remote system a certain degree of autonomy so that the "waiting gap" in the move-and-wait mode can be filled, as much as possible, with ongoing activities at the remote system, and commands for, and feedback from, every elementary action need not travel the communication distance between the human operator and the remote site.

The RMS under consideration here for the assessment of effects of communication distance consists of a planetary surface vehicle without and with partial autonomy connected by up-and-down radio communication links to an earth-based control center. The functions performed by the RMS are those required to accomplish the scientific objectives of the space mission. Generally these are to move the scientific payload on the remote vehicle safely from place to place, and to collect and report scientific information back to earth. These general functions can be broken down into seven specific system functions: (1) sensing of scientific and control data; (2) handling of samples and instruments; (3) controlling of scientific instruments and operational processes; (4) data reduction; (5) data displaying and recording; (6) appraising of control data, deciding and action initiating. and (7) scientific data interpreting (Ref. 5).

Certain assumptions about the relative importance of the various time lags can be made.

The two-way signal transit time depends only on the distance from earth to Mars and is in seconds approximately 6.67 times the distance in millions of kilometers.

The time to orient the imaging sensors will generally be short with respect to other times and can usually be ignored. Reading sensor data into storage depends on the type of sensors used, but even for TV sensor data, this time will be relatively short and can usually be ignored as well.

The time for on-board data processing depends heavily on the rover's automatic decision or data compression functions. If the rover does not perform any of these functions, the time for on-board data processing may be ignored. Otherwise it is reasonable to allow some time up to that required by ground-based data processing for trade-off analyses depending on the level of on-board data processing capabilities.

Antenna orientation time may be assumed to be inversely proportional to beamwidth. At the deep space communication frequency of 2295 MHz and reasonable remote antenna diameters between .6 to 1.2 meters the half power beamwidth varies approximately between 15 and 8 degrees, respectively. The time to orient the antenna will clearly also depend on the orientation technique.

The time to read sensor data into the telecommunications channel depends on the information rate of the telecommunication channel which in turn depends upon range transmitted power, vehicle antenna gain and the total bits transmitted per cycle. Earth reception is assumed to be through 64 meter diameter antennas. Within the parameter ranges of interest the data rate in bits per second can be closely approximated by the relationship $DR = 0.9\ PG\ (100/R)^2$, where P is the power in watts, G is absolute antenna gain and R is the distance between earth and Mars in millions of miles. The total number of bits to be transmitted is determined primarily by the video system which provides about 90 % to 98 % of these data. For preliminary trade-off analyses it is therefore sufficient to consider only video data, where the total number of bits transmitted is the product of the horizontal image lines, the vertical image lines, the number of gray level bits per element and a factor allowing for housekeeping (formatting, synchronizing, etc.). For a stereo pair of cameras the data is about twice as much and for colour it increased by a factor depending on the number of colour shades (usually three). The total data divided by data rate gives then the time to read the sensor data into the space communication channel. Also part of the communication channel is the ground communication system (GCS). The time to transmit data over the GCS from the deep space network receiving station to mission control at the space flight operations facility (SFOF) is negligible for the Goldstone receiving station which is linked to the SFOF by a wide-band channel. For overseas sites, however, the maximum bit rate capacity of the overseas network may cause a bottleneck. If the incoming data rate is greater than GCS capacity, storage must be provided as a buffer to handle the data queue with corresponding increase of time delay.

The time for ground-based data processing and control decision generation depends upon a great number of considerations not the least on human, managerial and organizational considerations discussed in the previous section. The time required for this function is most difficult to assess and may vary, based on latest Mariner Mars 1971 experience, anywhere between several minutes for the simplest decision cases to several days for decisions involving major strategy changes. This area is the least understood and requires still extensive investigations to assess the effects of this time delay on overall mission performance.

Read-out of commands into the GCS and up-link will be subject to the same kinds of information rate restrictions as the data transmitted on the downlink. The total bit content is much less, however, and the uplink data rate can be made much greater because of the substantially greater transmitter power.

The time for orienting the rover and/or the manipulators into the desired position may vary widely with the particular situation. Usually certain reasonable assumtions can be made however for performance assessment.

The time for actual locomotion is the stepdistance divided by the rover locomotion velocity. While rover locomotion velocity is dependent upon many parameters, it is reasonable to assume approximately 3 meters per second.

The overall average velocity of the rover is given by the stepdistance divided by the total cycle time, where the cycle time is the sum of all time delays enumerated above. A suitable measure for system performance is this overall average velocity. The performance of an RMS system consisting of the ground-based control center, the up-and-down communication link, and the remote rover will be better if the overall average velocity of the rover is higher. For a Martian surface rover this is, of course, not the only possible performance measure. In particular, as far as scientific analysis operations and the value of the returned scientific information is concerned, other performance measures could be established. These are, however, usually based on subjective and qualitative criteria.

Assuming that the roving vehicle on the Martian surface has in one case no on-board data processing capability and, in the other case, is equipped with so-called semi-autonomous capabilities, it is possible to perform comparative systems analyses for a set of representative assumptions including various antenna diameters, transmitted powers, numbers of vertical and horizontal elements for a stereo image system, and the like. Certain interesting points can be derived from examining the results of these analyses.

Without on-board data processing capability, only in very few extreme cases in the time spent in actual vehicle motion greater than 10 % of the total time. Usually it is around 5 % to 6 % . An increase in antenna gain is frequently a disadvantage, since the improvement in data rate is more than offset by the increased time for orientation. It is however not safe to draw firm conclusions regarding this aspect except to note that such an effect is in fact possible and should be carefully considered.

Increasing the transmitter power affects only the data rate and, hence, the time for read-out of sensor data into the telecommunication channel. Except for high TV resolution cases, there is little gained by increasing the power, and even then, the increase in overall average velocity is not much. Whether or not it is worthwhile to increase the transmitter power depends primarily on other trade-off system factors such as weight.

Since the video system is the primary contributor to the data to be transmitted, and since that data is directly proportional to horizontal and vertical lines of TV resolution, it is important to consider in any trade-off analysis what the effects of these two resolutions are. In most cases, it turns out that the maximum average velocity occurs with maximum vertical resolution, moderate horizontal resolution, high power and low gain antenna. The interaction of these effects can be explained by the fact that horizontal resolution is in general not as critical for the detection of an obstacle as is vertical resolution. For instance, the image of a crater at a certain distance, looked at by a camera mounted on the rover, appears as a horizontal ellipsis requiring numbers of horizontal and vertical TV resolution lines inversely proportional to the respective semi-axes of the ellipsis.

It is interesting to consider the relative effects of distance. The distance from earth to Mars varies from about 55 to 400 million kilometers depending on their relative position. This corresponds to a two-way communication transit time of about 366 to 2666 seconds. The general trend of the analysis results shows that as distance increases, a larger portion of the total cycle time tends to go into reading data into the communication channel and into spanning the distance itself. This then seems to be a major justification for going to a form of semi-automatic or supervisory control, where end goals and general path plans are transmitted to the vehicle from earth, but individual steering start and stop commands are generated by an on-board data processor directly responsive to sensors carried on the roving vehicle. The effect of this approach is discussed below.

The effect of data compression, whatever, its form of implementation, reduces the amount of data to be transmitted over the communication link in order to convey a given amount of useful information. For this, some penalty is paid in additional equipment at each end of the communication link and in slight time delays to process the data on board the rover. Taking these factors into account, the analyses show generally that data compression on the vehicle by a factor as small as 2 or 3 yield great performance dividends at maximum communication distances by reducing the time for readout of sensor data into the telecommunication channel; but at moderate distances compression factors as great as 10 have little effect on the performance.

For communication distances between 55 and 100 million kilometers, it appears that the time for ground data processing is by far the primary contribution to the overall control cycle time, while for larger communication distances up to 400 million kilometers the primary time requirement shifts to reading out sensor data into the communication channel and to signal transit. It does seem clear that at distances of 100 million kilometers and greater, it would be of progressively increasing advantage to provide substantial automatic capability to the rover in which on-board data processing and decision-making generate individual motion control commands for individual traverse steps to a destination via a route specified only in general terms by ground control. Comparison analyses have been made in which it was assumed, for instance, that ten such steps are made on the average between contacts with the earth-based control and that three stereo pairs are transmitted at the start of each cycle. The control cycle consists then of a minor and a major cycle, where the minor cycle is repeated ten times within each major cycle, and the minor cycle closes its control loop through the on-board computer of the rover, while the major cycle closes its control loop via the telecommunication up-and-down link through the human operator(s) on earth as shown in Figure 8.

The major conclusions from such comparative analyses are that substantial improvements in calculated performance with the semiautomatic mode of operation can be achieved for large communication distances. At 100 million kilometers distance the overall average velocity can be increased by at least 50 % , and in some instances as much as 70 %. At 400 million kilometers distance the improvement may go as high as 300 %, while the improvement at 55 million kilometers is not more than about 40 % . These improvements derive directly from the fact that time spent transmitting data from earth to Mars and vice versa has been reduced to a small portion of the total time. Actual locomotion time can be increased to approximately 20-25 % of the minor cycle, which in turn constitutes usually the bulk of the major cycle time.

VI. CONCLUSIONS

The programmatic environment foreseen for the late 70's and for the 80's within the U.S. space efforts is characterized by the availability of the Space Shuttle and Space Tug or other upper stage transportation systems. This capability will enable increased utilization of space for scientific pursuits and for the possible initiation of commercial ventures. One can foresee that space stations in earth orbit may function as remotely operated multipurpose scientific laboratories, e.g., as astromical observatories, as earth physics laboratories, etc., which will be used by scientists throughout the world much as are the Palomar Observatory, particle accelerators or the Magnet Laboratory here on earth, but which will be operated much of the time remotely from a control center on earth. In the future, space stations may also function as facilities for processing and manufacturing activities requiring a high vacuum and/or zero gravity environment. Such facilities could also be operated remotely from earth-based control centers, while the transportation requirements can be provided by the Space Shuttle and associated systems.

The operational Remotely Manned Systems described in this paper and the required technological capabilities will be necessary ingredients for the implementation of such possible future programs. In addition, a whole set of complex large-scale man-machine systems problems, of the kind identified on the basis of past space missions need study and resolution.

The primary purpose of the exploration RMS, in the forseeable future, is that of scientific information gathering. This has to be accomplished within acceptable mission times for several reasons. First, they have only a limited lifetime. Second, mission costs will increase with mission time primarily because of space flight operations. Third, the reliability of the system decreases with mission time. The control of spacecraft and vehicles from earth at lunar and planetary distances requires the development of techniques to deal with the problem of communication time delay so that the performance measured in terms of operations per unit time, or in terms of distance traveled per unit time, or in terms of any other suitable criteria can be increased to acceptable levels. In order to perform the required functions an extensive amount of scientific data and operational data for the purpose of moving from place to place on the Martian surface must be processed either on-board of the rover or on earth by machines and humans. A primary bottle-neck for the performance improvement of the system is the data rate capacity of the various elements of the Remotely Manned System consisting of the ground-based control center and the remote system.

This paper gives also a brief analyses of the times spent at various operations during a surface mission and identifies those parameters which contribute to the data rate and data processing limitations. Major results are discussed with respect to the important parameters of Martian missions and their performance measured in terms of the average velocity of a rover on the Martian surface. It can be concluded that the operational performance of a remotely controlled vehicle on the planetary surface depends primarily on the rate of useful information flow in the man-machine sensor-control loop. It also depends on the amount of automaticity or autonomous decision-making capability one can place on the remote rover system. This relieves the data flow requirement through the ground operation control center and in effect reallocates the decision-making process between the machine on the Martian surface and humans on earth. The machine performs the low level decision functions while humans on earth act as supervisors in the control loop.

Although these analyses were performed for vehicle motion and path planning, the results are indicative·also for manipulative tasks. It appears easier to breakdown manipulative tasks for well-defined experimental operations into pieces of task elements which lend themselves to supervisory control automation than it is to breakdown path planning, finding, and scheduling. The change from the move-and-wait mode to the supervisory control mode for manipulative operations appears, therefore, to be associated with even greater potential payoffs in terms of manipulation performance than for traverse operations.

REFERENCES

[1] Heer, E. , Ed., "Remotely Manned Systems — Exploration and Operation in Space," Proceedings of the First National Conference held at the California Institute of Technology on September 13-15, 1972; Published by California Institute of Technology, Pasadena, 1973.

[2] Johnsen, E.G. and Corliss, W.R., "Teleoperators and Human Augmentation, an AEC/NASA Technology Survey," NASA SP 5047, Dec. 1967; "Human Factors in Teleoperator Design and Operation," John Wiley & Sons, Inc., New York, N.Y., 1971.

[3] Deutsch, S. and Heer, E., "Manipulator Systems Extend Man's Capabilities in Space", Astronautics and Aeronautics, June 1972.

[4] "1973 Viking Voyage to Mars", The Viking Project Management, Astronautics and Aeronautics, Vol. 7, No. 11, pp. 30-59, Nov. 1969.

[5] Heer, E., "Remote Control of Planetary Surface Vehicles", 1973 IEEE INTERCON Conference Record of "Intercon 73" held in New York, March 26-29, 1973.

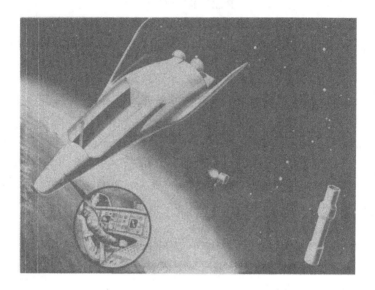

Fig. 1   Rendez-vous: artist's conception of a Space Shuttle-launched Free-Flying RMS performing rendez-vous with a satellite to be retrieved to the Space Shuttle.

Fig. 2   Space Tug: artist's conception of a manned Space Tug removing a space station module
from the Space Shuttle and mounting it to the Space Station.

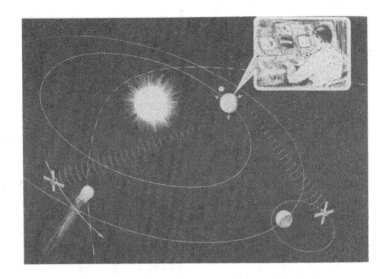

Fig. 3   Flyby and Orbiter Spacecraft: artist's conception of a Martian orbiter and comet flyby spacecraft envisioned as an RMS.

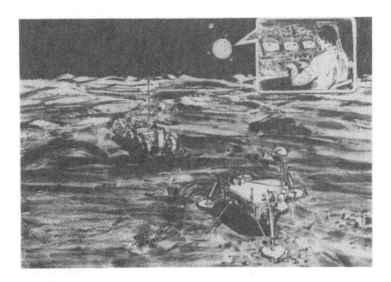

Fig. 4   Surface Explorers: artist's conception of stationary and roving extraterrestrial surface explorers envisioned as RMS. The stationary lander resembles the Viking 1975 Martian lander with sampling manipulator in the foreground. The small tethered rover serves for small excursions from the strationary lander while the larger surface rover may cover distances up to 1000 km.

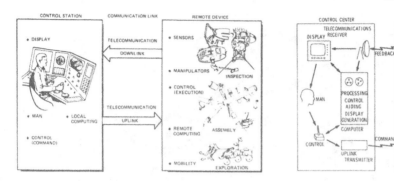

(a) RMS Concept   (b) RMS System elements

Fig. 5   Graphic description of the essential elements and interfaces of a Remotely Manned System.

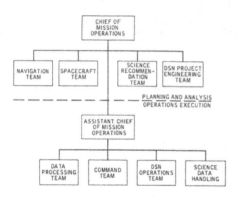

Fig. 6   "Human operator" organization of
Mission Operations System for Mariner Mars
1971, involving several hundred people in the
Control Center during peak of operations.

Fig. 7   Normalized task time versus total
time delay. Move-and-wait operator strategy
would be a succesful but slow strategy for
work on the moon and planets (ref. 2).

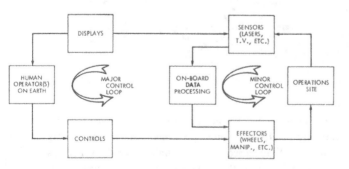

Fig. 8   Schematic of Minor and Major Control
Loops.

TABLE 1 CLASSIFICATION OF REMOTELY MANNED SYSTEMS (RMS)

RMS are man-machine systems that augment and extend man's sensory, manipulative and intellectual capabilities to remote places.		
Exploration RMS		**Operational RMS**
Tool for scientific exploration in space (information gathering)		Tool for engineering operations in space (technical servicing)
Flyby and Orbiter Spacecraft RMS		**Free-Flying RMS**
Examples	Mariner and Pioneer Planetary flyby Mariner Venus/Mercury mission Mars orbiter Jupiter orbiter Asteroid rendezvous Comet flyby	Examples — Control center is: In space station In space shuttle On the earth On the moon In space tug
Surface Explorer RMS		**Attached RMS**
Examples	Surveyor Lunar Lander Viking Mars Lander Lunar Rover Mars Rover	Examples — Manipulators are attached to: Space station Space shuttle Space tug

TABLE 2 INCREASING OPERATIONAL COMPLEXITY WITH COMMUNICATION DISTANCE

APPROXIMATE DISTANCE	COMPLEXITY	EXAMPLES
Within direct observation distance	Manipulator command and actuation precision Vehicle mobility performance	Exoskeletons ANL Model EI master slaves Earth surface vehicle Shuttle attached manipulators Space Station attached
Obscured position and outside direct observation distance to approximately 40 km	The above plus: Sensor performance Transmission of sensor data Information display	Shuttle attached manipulators Space Station attached Free flyers operated from Shuttle or Space Station Earth surface and ocean bottom vehicles operated remotely
Approximately 40 to 40,000 km	The above plus: Power bandwidth information rate tradeoffs Possible relay link requirements	Free flyers operated from earth-based control station
Approximately 400,000 km	The above plus: Spreading losses Transmission lags (seconds) Earth rotation Payload restrictions	Lunar orbiter Lunar rover Surveyor
Approximately 40×10^6 to 400×10^6 km	The above plus: Greater spreading losses Transmission lags (minutes) Relative motion of earth and planets	Mars orbiter/lander Mars rover Asteroid missions Venus/Mercury missions
Above 400×10^6 km	The above plus: Greater spreading losses Transmission lags (hours)	Jupiter orbiter and probes Outer planet missions

PROBLEMS IN SELECTION OF DESIGN PARAMETERS AFFECTING MANIPULATOR PERFORMANCE

Donald A. KUGATH & Donald R. WILT

Cybernetic Automation & Mechanization Systems Section
Re-entry & Environmental System Division
General Electric Company
Philadelphia, Pennsylvania, U.S.A.

(*)

Summary

The design and development of manipulators pose difficult cost/ performance trade-offs. Through its background and experience General Electric has developed a product line of Man-Mate® Industrial Manipulators. However, the published state of knowledge of manipulator performance is not broad or deep enough to provide a ready means for the designer to prescribe design parameters for a new or a special purpose manipulator, such as proposed for the NASA Space Shuttle.

This paper summarizes data in the areas of kinematics, compliance, and force feedback and relates these to manipulator design. Also, an approach useful in evaluating classes of manipulator for given applications and assigning figures of merit is discussed.

(*) All figures quoted in the text are at the end of the lecture

1.0. Introduction

The general Electric Company has been involved in the design, development and production of manipulative devices since the first modern innovations in manipulators, some thirty years ago. The manipulators began with radioactive "hot lab" applications and continued with special research devices such as Handyman, a hydraulic master/slave hot lab manipulator; the Quadruped transporter, a walking truck ; and Hardiman , the hydraulic powered "superman" exoskeleton device. Recently, these development programs have resulted in the commercial production of material handling manipulators called Man-Mate ® . The above devices are shown in Figure 1. Developmental work has also been done on converting the Man-Mate ® manipulator into the robot-like Computer-Mate.

This history of involvement in the technical evolution, plus constant sensing of the pulse of the other workers in the field in the U.S.A., has given General Electric a broad knowledge of manipulators. G.E. has considerable experience in the design and synthesis of robots and manipulator devices from the practical aspects. Having to design, build and sell actual commercial hardware results in very different design constraints than conceptual designs on paper.

There developed a state of the art, based perhaps too much on intuitive feel of how a manipulator should be designed. After all, there is not a large amount of published data in this area. There are, for example, only a handful of the 273 entries in the Bibliography of Johnsen and Corliss (Ref. 1) published in December 1967, that contain quatitative or comparative data that a manipulator engineer could use in his design development. There has been slight improvement since that time.

However, there are other reasons why good data on manipulative systems is difficult to find. A primary reason is the great dependence of task times on the feedback of information from the task site to the man operating the manipulator. The primary feedback loop is usually a visual link and many times this is the only feedback. We have found that task times are very sensitive to vision. Other factors such as friction, backlash,. compliance, torque/inertia, control location, and bias forces also account for definite changes in task performance times. Since there are no standard tests, in part because no society or organization has put forth evaluation schemes, the builder of a manipulator or robot evaluates the equipment against his assigned task, makes simple, easily implemented improvements where possible, and usually leaves it at that. The fortunate designer who can afford to build a variety of manipulator types or one device with variable parameters and evaluate in actual tests its capabilities to do the job presented is almost unheard of.

Considering the many task variables associated with manipulator applications in general, one soon reaches the conclusion that specifying a universal manipulator evaluation scheme is no trivial task in itself. The major task variables are tabulated below to help the reader appreciate the problem.

Manipulator Task Variables

1) Time
   a)   Work paced (e.g. assembly line or capture)
   b)   Self paced   (e.g. hot lab or disassembly)
2) Nature
   a)   Repetitive
   b)   Non-repetitive
3) Precision (accuracy of placement)
4) Dexterity (degrees of freedom required)
5) Unilateral vs. bilateral
6) Proximity (of operator to work place)
7) Load/reach ratio
8) Environment
9) Safety requirements

We feel that no one evaluation procedure can be used as a basis of comparison of all manipulators and yield meaningful results, unless we include cost effectiveness, which is the commercial manipulator buyer's primary concern. This also is a function of the task at hand. It is a measure of performance/unit cost and can be used to choose between manipulators. But even here the most cost effective may not be the best. If it is also for the most expensive, its pay-back period may be too long, resulting in a less costly, poorer performer taking first choice.

If the industrial manipulator is to succeed there must be a market and it must be a product that competes. An industrial manipulator is a high technology machine and as such is expensive. Therefore, it must have good performance to offset its cost by saving more money than it costs. In our efforts to enhance performance and reduce costs, we have studied the human factors and kinematics of our machines. Some of the findings are discussed in the following sections. A method of evaluating, in very general terms, the merit of manipulator performance is also discussed.

2.0. Kinematics

The designer must consider all aspects of the manipulator system and its application in choosing the kinematic arrangement. Prosthetic kinematics are prescribed by the limbs they replace. Space and submarine applications introduce stow envelope constraints in addition to problems associated with sealing of actuators, etc. In designing assembly line material handling devices, performance of the task takes precedence and is the driving design constraint. The constraints on the designer are more stringent when his work is going into a product. He has to consider component cost and reliability. In addition, general purpose manipulators must be readily adaptable to different applications. In larger sizes, the effects of systems power requirements and weight gain in importance.

Initally, G.E. built a 150-pound capacity manipulator, which had an articulated arm with shoulder and elbow joints in series on a base that rotates in azimuth as shown in Figure 2a. In this arrangement, the torque at A is also applied at B. If we neglect the weight, the joint torques may be expressed as:

$$T_A = WL \cos (\theta_A + \theta_B)$$

(1)

$$T_B = T_A + WL \cos \theta B$$

To supply these torques with linear hydraulic actuators, which generate essentially sinusoidal torque curves, results in considerable inefficiency. This is illustrated in Figure 2b.

As the business grew, a larger capacity manipulator was designed, which uses a parallel arrangement, shown in Figure 2c. This mechanism transmits the torque at A as an axial load witout exerting a torque about B, therefore, a smaller actuator can be used at B. Another benefit of the parallel arrangement is that the actuator at A is now in a fixed reference system. Because of this, its torque can be phased with the load torque, which results in actuator A being smaller also. The equations for the load torques in this case are:

$$T_A = WL \cos \theta_A$$

(2)

$$T_B = WL \cos \theta_B$$

The torque curves are plotted in Figure 2d. From which we find that a 40% reduction in joint torques is possible, a good practical application of kinematics. This kinematic arrangement is used in the G.E. Man-Mate CAM 1600.

There are other benefits of the parallel link arrangement. Referring again to Figure 2a and 2c, note that with the parallel system the joint motions are closer to orthogonal. This has benefits from the control standpoint, particularly in the case of a robot, where orthogonal motions simplify coordinate transformation and minimize perturbations in adjacent joints due to compliance.

Straight line mechanisms have been used in shipyard level-luffing cranes for many years. Adapting them to our manipulators was the next step toward improving system performance. Figure 2e shows the kinematic arrangement for the new G.E. manipulator (4000 lb. capacity). This machine is a further advancement, providing improved efficiency, more payload per unit weight and has the added benefit of a compact stow envelope. It accomplishes all this with an approximate straight line motion which reduces the torque at B. The result is a small increase in torque at A, with a large decrease in torque at B. This is illustrated by the torque curves plotted in Figure 2f. The net savings in power is approximately 35 % , compared to the parallel link mechanism. The power savings for the actual machine, however, which has link weights and acceleration forces to contend with, is less than 35 % , but is still large.

The evolution of the general purpose manipulator kinematics has been driven primarily by economics as size increased. The benefits of improved performance realized can now be factored into all future designs, large or small.

3.0. Compliance

In space applications such as the Shuttle and Space Station, long reach manipulator-boom systems have been proposed to do tasks of assembly, docking, deployment, and inspection. Booms up to 60 feet in length have been suggested (Ref. 2 and 3). Because of their length and the need for light weight construction, they are more flexible or compliant than present earth-based manipulators. As a rule, designers like to make manipulators stiff, not only the structure, but also the servo.

In 1972 NASA/JSC funded General Electric to participate in a joint study to determine experimentally the effect of compliance on manipulator performance (Ref. 4). Two tasks of different nature were each performed on two different manipulator systems with servo compliance as the dependent variable. The first task was tracking, in which the manipulator guided a large mass supported on an air bearing floor through a two-dimensional maze. The second task consisted of bringing a relatively large, moving mass to rest in as short a time as possible. The first manipulator was an E-2 model electromechanical system. This is a bilateral, man-equivalent manipulator with a 6-pound force capability. The upper arm is 18 inches and the forearm is 30 inches in length. Although it has 6 degrees of freedom, only the three transport motions were used in the study. The second manipulator had an electro hydraulic system. It was a modified General Electric CAM 1400 model, industrial manipulator. The modifications were: forearm and upper arm lengths, increased to 13x12 feet; master station moved for remote operation (normally the operator rides in azimuth with the boom); and adding a servo pitch end effector motion which maintains a fixed attitude relative to ground.

3.1. Description and Results of Compliance Experiments

The two tasks envisioned were mass catching and mass positioning. The former would be a work paced task analogous to the capture of satellites in space, while the latter would be analogous to positioning of payloads in and out of the shuttle at the operator's own pace.

The subjects used in all the tests were NASA/JSC engineers involved in the space manipulator program.

3.1.1. E-2 Pendulum Catching.

The use of a 33-foot long pendulum with its easily controlled energy and path precluded variations due to different amounts of energy to be absorbed or task time variations because of differences in the subjects catching ability. Each subject caught the same 400-pound mass at the same position (at the maximum displacement of 24 inches) and brought it to rest as quickly as possible. The dependent variable was servo compliance, which was changed in four steps over a range of an order of magnitude. Data was collected by a strip chart recorder which showed the pendulum position and E-2 motor voltages as a function of time.

The E-2 pendulum test showed an increase in task time with increasing compliance. However, the task time only changed from 4.64 seconds to 5.47 seconds as the compliance, measured at the wrist point, increased from 15 in/lb. to 1.8 in/lbs. This is a small percentage (18 %) increase in time, but if one considers that 4.64 sec. approaches the theoretical minimum time (based on a maximum given E-2 force level) the increase is more significant.

3.1.2. E-2 Maze Test

A 650-pound mass on air pads was guided through a maze path about 48 inches in length. The E-2 gripped the mass above the c.g. on a 5/8 inch O.D. pipe which was the path follower and also the air supply. The width of the maze slot was 1.50 inches, giving a clearance of 7/16 inch on each side. The subject sat immediately next to the maze and directed the movement of the mass by applying the appropriate master motions. Originally, it was intended to have the subjects restrain their speed and avoid all contact. During initial tests, the learning curve was rather lengthy and data not consistent enough. The subjects were then told to go as fast as possible yet avoid contact, i.e., the emphasis shifted from accuracy to speed. The data consisted to the task time and number of errors (wall contact).

The E-2 maze test results are clouded by the fact that apparently not enough data was taken. The subjects were still learning in many cases. However, analysis of the data predicted longer times with increase in compliance when learning would have been complete. The increase, however, would be small. The percent increase was less than that of the pendulum test. Again, the subjects came close to the theoretical minimum task time.

3.1.3. CAM 1400 Tests

The maze test was repeated but at a 4:1 scale up in maze size and an increase in mass to 7,000 pounds. The compliance could not be varied as greatly as with the E-2, and only two values were used. Since enough data points were not taken with the E-2 maze test, during this test the full learning curve was explored.

The CAM 1400 maze test data was sufficient to show definite results. The two compliance settings, a 5:1 change, had differences in time of slightly more than 1/2 second. Low compliance runs averaged 17.91 seconds in the clockwise direction and 16.61 seconds in the counter-clockwise direction while, the higher compliance runs averaged 18.46 and 17.23 seconds. Thus, the change in direction was more significant than the change in compliance. For some of the subjects, it is difficult to tell from the learning curve plots when the compliance changes were made. Learning curves for two subjects are shown in Figure 8.

3.2. Conclusions on Effects of Compliance Changes.

The overall conclusion of the Nasa/JSC data is that large changes in compliance need not appreciably change task times for these types of tasks with a bilateral manipulator. In these tests, only the servo compliance was varied. The effect of structural compliance increase would have the same effect provided the system is well damped. Thus, it is suggested to provide damping for long, slender structures, such as the space manipulator boom. It is significant that the task times approach the theoretical minimum times, given enough practice. The small increase in task time with compliance are due mainly to the increased motion of the master to achieve optimal (i.e. maximum) force levels.

4.0. Force-Feedback in Manipulator Systems

Some of the earliest types of manipulators had force-feedback, since they had direct mechanical coupling. With the desire for more mobility, the use of motor driven units, which did not have force-feedback, evolved. Relatively few systems have been built with bilateral (force-feedback master/slave) control. However, a good share of these bilateral systems, and particularly in high gain systems, were developed by General Electric. The Handyman manipulator had force-feedback incorporated in all ten joints. Built in 1954, this manipulator has an exoskeleton master arrangement from the forearm out to the fingers. Being complex, and therefore expensive, plus having been developed under a government program which was terminated, Handyman's development was not pursued further. The Quadruped walking truck has twelve force-feedback motions, three in each of the four legs. Force-feedback proved to be a necessary feature. Without it, attempts to walk the truck were futile. Following the Quadruped in development was the Hardiman project. This exoskeleton device has 30 servo joints. The general purpose boom manipulator developed by General Electric also used force-feedback in the three arm (transport) joints, but so far have not used bilateral control in the wrist (orientation) joints. All of these devices used hydraulic power systems. We have also built electromechanical joints with force-feedback and have a Model E-2 manipulator in our laboratory. The E-2 is very useful in force-feedback studies because of the ease with which the force-feedback can be switched on and off.

4.1. Effects of Using Force-Feedback in Manipulators

Based on our experience with building bilateral systems and conducting test programs evaluating the usefulness of force-feedback, we feel we have a good understanding of the subject. The benefits of using force-feedback will be discussed first, followed by the disadvantages and problems associated with it.

One benefit is the compliance added to the slave portion of the system. Since the slave load is sensed, and a proportional signal sent back to the master immediately, better control can be accomplished. The compliance is due to "give" of the operator on the control and also because his reaction time is considerably faster when he feels a force pushing in the proper direction than when he has to see what is happening at the slave and mentally calculate the proper control input. Secondly, by using a bilateral control, the operator can more effectively control the slace, especially when the slave is larger than the master. Since the operator must control the reflected inertia and motion of the slave, the control acts as if it is overdamped, while the same manipulator without force-feedback tends to overshoot. Thirdly, force-feedback allows slave/master position ratios much greater than one. With unilateral control, it has rightly been said that using high slave/master ratios would produce a jittery slave, i.e. every little motion or twitch of the man operating the master would be magnified in the slave. However, with bilateral control, all jitters are effectively filtered out, yet allowing small motions of the slave if desired. The smoothness of performance reduces task time variations, not only in boom type manipulators doing material handling tasks, but also in man-equivalent tasks. Fourth, task capabilities are increased significantly in most types of tasks. For example, the inertia of the load can be better sensed (i.e. acceleration feedback in addition to the velocity and position sensing of the operator) and the proper forces better introduced to the load. In fine positioning or low inertia tasks, times are better when contact forces are part of the task. If contact is used as an aid to positioning (e.g. guide pins or guide rails), force-feedback will sense contact. And, when contact is inherently required, force-feedback will prevent the contact forces from becoming too large. Finally, related to the last point, force-feedback allows reduced power consumption when comparing a bilateral position control to the equivalent unilateral control, since forces greater than needed are not used.

Of course, force-feedback comes at a price. The added complexity will necessitate a higher production cost. Added components could potentially add to reliability problems. However, by use of a bilateral control, reliability could actually be increased over an equivalent non-force-feedback design for certain tasks since forces can be better controlled. The volume requirements at the master are greater for the bilateral control, especially when compared to the simple on-off switch control or sophisticated hand controllers. The added force-feedback actuators also add to the master weight. Both volume and weight are important in such application as space in underwater. Finally the forces felt by the operator add to his fatigue problem. We have kept the forces down to the 5 to 10 lb. level to minimize this problem. Even with the problems associated with force-feedback, sometimes the task can only be done by a bilateral manipulator, or alternately be done so much better that the added cost would not be a driving factor.

4.2. Experiments Involving Force-Feedback

The generalizations above on force-feedback are gathered from experimental data, as well as experience. Some of the more pertinent test results will be summarized below.

4.2.1. Experiments With the E-2 Manipulator

The model E-2 electromechanical manipulator's (shown in Figure 4) master is identical to the slave, except for the terminal device. The E-2 can presently lift six pounds and can move at greater speeds than 30 inch per second. By switching off the fixed field of the two-phase AC servo motor on the master, the unilateral mode can be studied, keeping force capabilities, kinematics, speeds, etc. equal to the force-feedback mode. Three different experiments of the many tests run will be reported below.

4.2.1.1. Peg-in-the-Hole Experiments

The details of this series of tests have been reported (Ref. 5) but, because of their significance, the results are summarized here. As shown in Figure 9, the task was a simple transfer of square pegs into holes with controlled clearances. The tests here used 0.01 inch clearance on each dimension of the 1.5-inch square hole. The use of force-feedback clearly reduced average task time by nearly a factor of two. The variability of the task times was considerably smaller for force-feedback and the learning curves smoother. It was found that the task time ratio of force-feedback to non-force-feedback was about constant as vision degraded either by increased subject to work site distance or by the use of closed circuit TV for visual feedback. Thus, there was not a constant time factor required by not using force feedback. It was also found that the average power consumption of the manipulator increased by a factor of three when force-feedback was not used. This is due to high contact forces the operator was unaware he was using. Coupled with the almost doubled task time, this results in a factor of six increase in total energy to perform this task without force-feedback.

4.2.1.2. Assembly Task With Vision Degraded

This task consisted of inserting a flanged tube into a hole, twisting the tube until an alignment pin and hole mated, and then inserting, but not torquing bolts. The task is shown in Figure 5. The tube diameter of 2-inches will indicate relative size. Visual feedback was by use of a single camera closed-circuit TV system. The vision was degraded by reducing the light illumination. Because of the single camera without pan and tilt, an assistant stood neat the work site to insert the magnetic socket wrench and bolts at the appropriate time. The time to manually load or unload the tongs with the bolts, wrench, etc. is not recorded as part of the task time. The task time shown below is for both removal and replacement.

Four conditions of vision were tested, rated as good, fair, poor or no vision at all. It was found impossible to do the task without force-feedback and without vision, whereas, with force-feedback, it was possible and times could be substantially improved when an auxiliary aid was used at the master. This aid consisted of an adjustable grid work of the same dimensions as the work piece. By grasping the work piece and poking around until an edge was felt, aligning the appropriate grid line with that point, going on to the next edge, etc., the grid could be aligned at the master to represent the task at the slave. By knowning the relative position of the holes to that grid, and interpreting the blind tapping force-feedback signals, the hold could be found and the tube or bolts aligned for insertion. Obviously, this technique works only because the work or task dimensions were well known.

The table below summarizes the result for one trained operator. Note here that as vision degrades, the task time ratio increases, in contrast to the nearly constant ratio of 2.0 for the Peg-in-the-Hole task. The number of fumbles, i.e., dropped work item or bolts is significantly greater without force-feedback. This

Visibility Condition	Average Task Time & Fumbles With FFB	Without FFB	Task Time Ratio	No. of Trials
Good	88 sec. 0	122 sec. 1	1.4	10
Fair	130 sec. 0	236 sec. 3	1.8	10
Poor	251 sec. 0	573 sec. 7	2.3	10
Blind	1184 sec. 0			

follows the trend of previous experiments. The errors are mainly due to lack of feel in the grip. The time lost in retrieving the work items or bolts is not recorded as part of the task times, and thus actually, the task time ratio would be greater.

4.2.2. Experiments Involving the Man-Mate Industrial Manipulator

The Man-Mate Industrial Manipulator (see Figure 1), built by General Electric for a wide variety of material handling tasks, incorporates as a standard design feature force-feedback in the three arm joints, azimuth, shoulder and elbow. Two machines have been modified to run without force-feedback and a number of experiments run. The boom comes in three models, which vary in size and load capability, but are all quite different from the E-2. Load ratings run from 100 to 1,000 pounds, with slave booms from 5' x 5' in length (forearm x upper arm) to 12' x 13'.

4.2.2.1. NASA/JSC Experiment Without Force-Feedback

The test set-up and description of the Model 1400 Manipulator have been described above in the section discussing compliance. In the maze test, all the subjects has successful (i.e. no touches) runs, using force-feedback, with times between 12 and 20 seconds. However, with the force feedback disconnected, the task without any touches seemed impossible. Over forty recorded runs and numerous practice runs were made, but none without errors. Most of the early runs had rather serious errors, such as moving the whole maze several inches or even knocking it down. The least number of errors in a recorded run was three, but the average number of touches was 9.5. Some operators learned from experience to use two hands on the master for a steadying effect. Even with the 7,000-pound mass at the end of the boom, it was difficult not to put in unwanted or unexpected error signals, due to hand jitter. Additionally, there was the problem of keeping the master in the proper plane, which matched that of the air bearing floor. If the master was raised or lowered significantly, the boom up or down force could exceed the grip force and the jaws would slide up or down on the pipe. Most operators learned to overcome this latter problem to a great extent by watching the deflection of the end effector in pitch and using the visual feedback to correct the up or down error. The jitter problem caused by the sensitivity of the boom's high gain in position was never completely solved, since no operator to date has been able to keep the pipe within the track — no matter how slowly and carefully he went.

To test the capabilities of capturing a moving object, an experiments was run in which the operator tried to catch the 7,000-pound mass moving by at various speeds. At the time, there was no precise energy delivery available for moving the mass. Thus, two men were used to push it. The test conductor determined its velocity by measuring the time to travel between two lines on the floor and the command given to the boom operator to start. The operator would then rotate in azimuth about 50° CCW, then track the catch point, while adjusting the yaw motion and closing the jaws.

The results for this test are summarized in the bar charts of Figure 6. The graphs show the number of successful and unsuccessful captures as a function of velocity for the full force-feedback and no force-feedback. At higher velocities, because of lack of tracking time and the quicker movements required, it is expected that the relative number of successes should decrease. The data shown is for a number of experienced boom operators, although the runs above 2.0 fps were done by only oneoperator, who was better than average.

For the non-force-feedback case, although now the task was possible , comparison of the bar graphs shows the reduced velocity capabilities due to the lack of force-feedback. Below 1 fps, the success ratio is only 53%, compared to 100% for the feedback mode. The greatest speed of a successful capture without force-feedback is 1.3 fps, compared to successes at up to 2.7 fps with feedback.

The conclusion of the experiment was that the lack of feel for the inertia of the boom did not allow precise tracking and adjustement, and thus made the task considerably more difficult to not only line up and capture, but also caused harder captures. That is, with the force-feedback, the mass trajectory was rather smooth, but without force-feedback the mass could be hit harder, and thus cause a more discontinuous path of motion.

4.2.2.2. Hole-Poking-Experiments

After some thought on how to evaluate the positioning capabilities of the Man-Mate booms when they were in production test, a hole poking experiment was devised. The test procedure and apparatus are simple, and simulate a portion of the actual work cycle of the manipulator. The apparatus consists of two targets, about one foot square, which can be moved to various points in the working volume independently. Touching the start target initiates a timer, and when the stop target is hit, the time is recorded to 1/10 of a second accuracy. The stop target has a removable face which has various size slots cut into it. The slots are all 4-1/2 inches high, but vary in width. A probe is attached to the end of the manipulator, which passes through the slot about 2 inches to turn off the timer. Thus, both the distances traveled and the final boom positional accuracy are controlled by the experimenter.

Tests were run in General Electric's labs to determine the effect of force-feedback, using the experimental set-up. One of our most skilled subjects performed the test using a unit on which force-feedback can be removed. The comparison of times is shown below. It was evident that learning took longer without FFB, and therefore more trials were needed. Besides the increase in task time, the variation was much greater. This variation is equally important as mean task times for prediction of task cycle times which the manipulator has to meet.

	Nbr. of Trials	Mean Time*	Std. Dev.*	Best Time	Worst Time
With FFB	60	2.33 sec.	0.21 sec.	1.8 sec.	2.8 sec.
Without FFB	90	3.32 sec.	0.77 sec.	2.2 sec.	5.3 sec.

* Based on last 30 trials

5.0. Figures of Merit

The previous sections discussed three manipulator parameters that the designer/user should consider. The effect on performance of these parameters has met sparse treatment in the literature. Since it is not practicable to test experimentally the hardware for each new application, it would be useful to develop figures of merit to provide a means of analytical synthesis of a manipulator. Such figures of merit would serve three purposes: (1) they would be used as a measure of performance to evaluate design variations, (2) as a means of comparing a new machine's performance to that of an accepted standard (i.e. quality control) and (3) to compare performance of dissimilar or competing manipulators.

As an example, one figure of merit we have found to be useful is the task time for the "poke-in-the-hole" test discussed earlier. Vision and learning are two significant variables, the effects of which can be controlled in this test. Since it is a simple task that can be rapidly learned, it provides an easy, fast means of evaluating the learning curve. Depending on the placement of the targets, individual motions or combinations of motions may be tested. Varying the size of the stop target slot provides a means of measuring position accuracy.

To account for the effect of both distance between targets and accuracy, we make use of an index of task difficulty first proposed by Fitts (Ref. 6) and later by Ferrel (ref. 7) for a two degrees of freedom manipulator experiment.

Defining the index, I, as :

$$I = \text{Log}_2 \frac{2 \text{ X distance moved}}{\text{final clearance}}$$

the index for the "poke-in-the-hole" test described in Section 5 would be :

$$I = \text{Log}_2 \frac{2 \text{ X } 100}{4} = \text{Log}_2 50 = 5.65$$

Tests performed with the Man-Mate ® manipulator indicate that, within limits, as I increases the task time increases nearly linearly, i.e.,

$$t = a + b I$$

When $I = 1$, $t = a + b$ operator's reaction time. Our tests have verified this result. The slope, b, is a function of the machine parameters (e.g. inertia, friction, servo performance, kinematics, speed, etc.).

For large values of I, the equation becomes non-linear. This may be attributed to an extremely small position tolerance or to a large transport distance. The latter has its effect due to the velocity saturating for a sizable portion of the task time. However, by graphing t vs I a comparison may be made in the non-linear regions also.

In Section 5, we used the "poke-in-the-hole" test to evaluate the effect of force-feedback. The test may also be used for comparing different manipulators. For example, a test was conducted which compared the Man-Mate® to a model M-8 hot lab manipulator and to a man for the case when $I = 5.5$. The results of this test are tabulated below.

	Mean Time	Std. Deviation	Hindrance
Man	0.67	0.067	1.0
M–8	1.20	0.079	1.8
Man-Mate ®	1.78	0.166	2.6

The manipulators are compared by means of the hindrance, or ratio of task time to that required by a man for a similar task. This comparison did not take into account the load capability. Had the man, for example, been required to move a 100-pound load, then the hindrance factor for the Man-Mate ® would very likely be much less than one.

The "poke-in-the-hole" test has thus far only been used to evaluate transport motions. It could be adapted to evaluate the wrist or orientation motions also. We are embarking on a study to develop such a test. Initially, our goal is to relate these motions to the index I such that a proposed repetitive task may be analyzed and the cycle time predicted.

There are many other figures of merit that, due to limited space, we cannot treat in any detail. A list of some key figures of merit is given below.

lift/weight	learning curve
working volume/machine volume	friction
lift x reach/power	fatigue level
lift	remote operating capability
reach	environment capability
power	cost (first cost (operating & maintenance
dexterity	reliability
response	availability
precision	safety (fail safe).

This is not a complete list and the reader may have items to add from his own experience. We have presented herein data and ideas from our own experience with the wish to augment the technology through stimulating discussion. The hope being that it will further understanding of these complex machines and their applications.

REFERENCES

[1] Johnson, E.G.; and Corliss, W.R.:"Teleoperators and Human Augmentation„ NASA SP—5047, Dec. 1967.

[2] Martin Marietta Corp.:"Preliminary Design of a Shuttle Docking and Cargo Handling System„ Final Report NASA/JSC 05218, Dec. 1971.

[3] MB Associates:"A Shuttle and Space Station Manipulator System„ Final Report NASA/JSC 05219, Jan. 1972.

[4] General Electric Co.: "Experiments Evaluating Compliance and Force Feedback Effect on Manipulator Performance„ Final Report NASA/ JSC 07239, Aug. 1972.

[5] Kugath, D.A.; et al:"A Remote Manipulator System for Space Applications„ Proceedings of International Symposium on Man-Machine Systems, Cambridge, England, Sep. 1969.

[6] Fitts, P.M.:"The Information Capacity of the Human Motor Systems in Controlling Amplitude of Movement„ J. Exp. Psy. Vol. 67, 1964, pp. 103-112.

[7] Ferrell, W.R.: "Remote Manipulation with Transmission Delay„ NASA TND-2665, Feb. 1965.

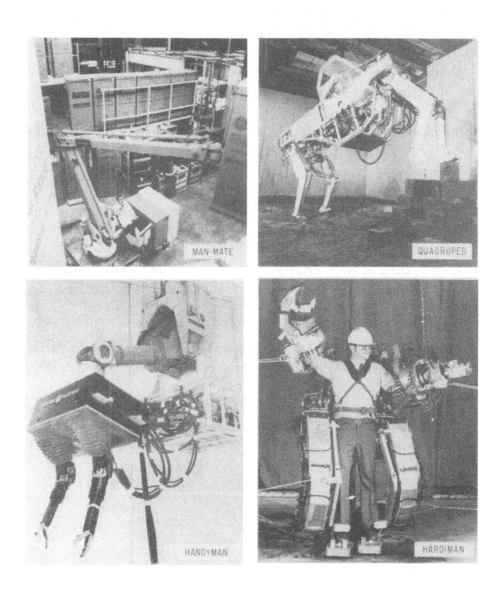

Fig. 1 Manipulator systems developed by General Electric

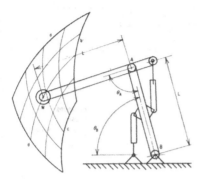

Fig. 2a Series design

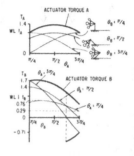

Fig. 2b Series torque curves

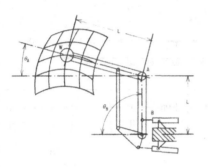

Fig. 2c Parallel design

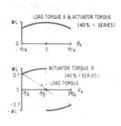

Fig. 2d Parallel torque curves

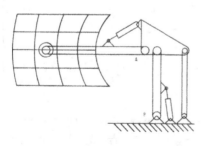

Fig. 2e Straight line design

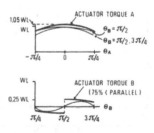

Fig. 2f Straight line design

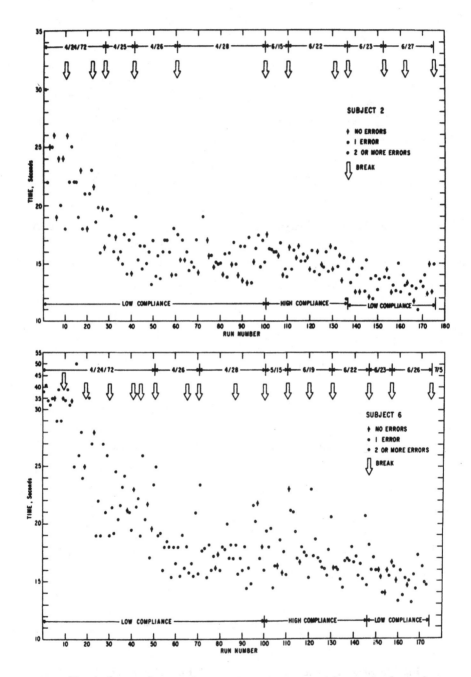

Fig. 3 Effect of compliance changes in maze test with manipulator boom

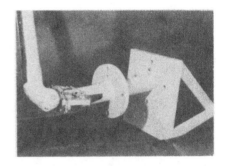

Fig. 4 Peg-in-hole task   Fig. 5 Assembly task

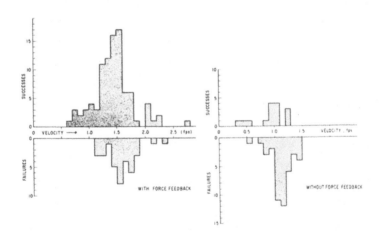

Fig. 6 Large mass capture with manipulator boom

CRITERIA FOR EVALUATION OF KINEMATIC CHARACTERISTICS OF MASTER-SLAVE MANIPULATORS AND A METHOD FOR THE STUDY OF THEIR SPACE PERFORMANCE

Gleb I.LUKISHOV, Chief of the Department,
State Designing Institute of the Soviet Union,
State Committee on Atomic Energy, Moscow, USSR

Yuri V. MILOSERDIN, Professor,
Moscow Engineering Physics Institute,
Moscow, USSR

(*)

РЕЗЮМЕ

Предлагается ряд критерев для оценки кинематических качеств копирующих манипуляторов, с помощью которых можно проводить качественное и количественное сравнение зон обслуживания манипуляторами различных конструкций, и дается метод исследования их пространственных характеристик. Знание характера изменения площадей и объемов обслуживания манипуляторами позволит рационально проектировать рабочие места установок с применением копирующих манипуляторов.

(*) All figures quoted in the text are at the end of the lecture.

1. Introduction

The most distinguishing feature of master-slave manipulators is that their master and slave arm designs [1,2] are kinematically alike due to which the motions performed by the master arm are duplicated by the slave arm. In this case it should be kept in mind that the motion duplication may be of direct or reverse action as well as reduced or non-reduced. As a rule, in such manipulators provision is made for the feedback of feel which ensures good sensibility of the operator's hand to loads.

The master-slave manipulators are most conveniently used to perform precision work in a comparatively small volume within the limits of spatial movements of the operator's hands. Such manipulators show great maneuverability, quick action and convenience of control as the operator working with the manipulator performs quite familiar actions. In his work the operator automatically sets up a programme of motions correcting them through visual observation and sense of force reflection.

It is advisable to design such manipulators only for low load capacities (not more than 25 kgf), because large loads develop substantial inertial forces severely decreasing working speeds of movements. Besides if the operator has to receive forces of more than 5 kgf he gets easily tired and accuracy of the work performed suffers. Therefore, even in manipulators with a load capacity of more than 10 kgf provision shall be made in principal diagrams for reducing the forces in their transmission from the slave arm to the master arm.

To date master-slave manipulators are most extensively used. For load capacities up to 5 kgf direct-acting (mechanical) master-slave manipulators are mainly used, and for load capacities of more than 5 kgf - manipulators of reverse action are mostly used (e.g. electro-mechanical type).

In their construction the well-known models of master-slave manipulators may be divided into three groups, namely: articulated tongs, telescopic and articulated manipulators [3]. All these manipulators are versatile in operation and have sufficient degrees of freedom to carry out remotely rather complicated work. Presently there are a number of manipulators [4,5] differing from one another in principal and kinematic diagrams and in their geometrical size as well.

To evaluate the manipulator kinematic characteristics qualitatively appropriate quantitative data on their space performance shall be established.

2. Selection of Manipulator Kinematics Criteria

It seems to be technically justified to use absolute as well as relative parameters showing the space potentialities of the manipulator tongs as criteria for evaluation of manipulator kinematic characteristics.

Absolute parameters include the area served by a manipulator and the area served jointly by two manipulators.

The manipulator service area is a space in which tongs may occur at any point in the course of all possible motions performed by the master arm.

The area served by a master-slave manipulator is considerably limited by the operator's anthropometric data. The methods are used now to enlarge this area, they are:

— reduction of the length of slave arm kinematic units relative to the master arm,

— introduction of additional adjustment motions for the slave arm.

The area jointly served by two manipulators is that part of the space in which tongs of both manipulators may be simultaneous. This area must be most comfortable to allow for simultaneous work with two manipulators.

Absolute parameters can also include the areas covered by manipulators at various levels.

The absolute parameters allow the evaluation of the working volume in which a manipulator may be used. However they do not show at what expense (size of additional volumes used) it is achieved. Therefore there is a need for such an index which would represent the ratio of the volume seved by a manipulator to the total volume occupied by the slave arm in operation. This index is particularly important because any increase of the working volume (e.g. box, cell) involves additional capital costs and a large-sized slave arm of the manipulator with a small service area will demonstrate its low-efficient kinematic diagram. The ratio of the whole volume served by the manipulator to the volume occupied by the slave arm in operation may be called a kinematic coefficient of the manipulator (K_M). The volume occupied by the slave arm is suggested to be defined as a rectangular parallelepiped circumscribed around the extreme positions of the structural elements during operation throughout the service area.

The rectangular parallelepiped is adopted because the majority of the box and cell constructions are of the same shape or similar to it.

The kinematic coefficient of the manipulator shall be considered for one as well as for two jointly operating manipulators and shall be determined from the following relationships:

$$K_M = \frac{1}{V_o} \int_v f\ (x,y,z)\ dv\ ;$$

$$K_M = \frac{1}{V_o} \left[2 \int_v f\ (x,y,z)dv\ -\ V_c \right]\ ;$$

where

v_o – volume of the parallelepiped circumscribed for one manipulator,

V_o – volume of the parallelepiped circumscribed for two jointly operating manipulators,

V_c – volume jointly served by both manipulators,

Operation in different points of the manipulator service area is not equally convenient; that is why it is necessary to study the maneuverability of the tongs in all the points of the volume. Thus, it does matter by which means the tongs can approach the service point - with limitations or without them.

As a criterion characterizing the manipulator performance in regard to its tong angular orientation freedom the report [6] suggests use of a parameter which is a mean value of the service coefficient $\bar{\Theta}$ in the working volume (v) of the manipulator service area. It is determined from the following expression:

$$\bar{\Theta} = \frac{1}{V} \int_v \Theta dv\ ;$$

But

$$\Theta = \frac{\Psi}{4\pi}\ ;$$

where

Θ – a service coefficient at a given point,

Ψ – a service angle that is a solid angle within which the tong can reach this point.

The value of the service coefficient at a given point, may vary from 0 for points at the boundary of the manipulator service area to 1 for complete service points. However for articulated tongs and telescopic manipulators of standard models there are no complete service points in the area served by one manipulator and maximum value of the service coefficient does not exceed 0, 4 ÷ 0,7.

3. A Method for Studying Manipulator Space Performance

As mentioned above, the manipulator coverage is the most important kinematic characteristic of a manipulator. However this characteristic is usually not determined for every manipulator model but is prescribed indirectly in terms of maximum coordinate and angular motions of the manipulator slave arm head. This information is not sufficient. It does not establish the mode of the manipulator service area changes in horizontal sections over the entire height of the manipulator coverage with due regard for the anthropometric data when the operator is in his normal position.

Thus the establishment of the actual manipulator coverage is a necessary ground for the evaluation of the manipulator potentialities at work as well as for the selection of the optimum working volume dimensions of the installations with manipulators (cells, boxes etc.). Understanding the mode of the service area changes in horizontal sections is very important for the selection of the optimum position of the main work plane (top plate) relative to the manipulator supports. Therefore the determination of the region of maximum service area is also a practical necessity.

To solve these problems a method of volumes for studying the manipulator space performance is suggested, which incorporates the construction of the service area diagrams and the establishment of the service area and volume dependance on the height of the horizontal section plane.

The results of the study performed for articulated tongs with a wrist unit are shown in the figure.

The investigations were carried out in accordance with the procedure which comprises the following stages:

a) The manipulator skeleton diagram is drawn to scale in two projections: a longitudinal view and a front view.

b) The extreme positions of the manipulator slave arm when the operator is in normal position relative to the master arm are marked on both projections.

c) The volume covered by the manipulator is divided by horizontal planes at 50-100 mm intervals. Each plane is designated by a numerical index. The plane passing through the manipulator support is the datum plane for counting; it is indexed by "0" (zero). 50 mm and 100 mm intervals are chosen as the most convenient to work with. One should understand that the smaller the interval module, the more accurate the results of the analysis, but the investigation becomes more labour-consuming in this case.

d) Based on the manipulator coverage sections the manipulator service area diagram is constructed for one manipulator as well as for two manipulators operating jointly, the section outlines being designated by the indices of the corresponding horizontal planes. Of course, only those section planes which pass through the manipulator coverage are shown in the diagram. Their values are put down in the table.

e) The table data are used for plotting areas served by one or two manipulators, $f_M(H)$ and $F_M(H)$, as well as areas served by both manipulators operating jointly, $F_c(H)$, against the height of the section plane.

f) The volumes served by one or two manipulators $v_M(H)$ and $V_M(H)$, as well as volumes served by both manipulators operating jointly, $V_c(H)$, are plotted against the height of the section plane.

g) The boundary lines (see dotted lines) of the areas f_o, F_o and volumes v_o, V_o of the parallelepiped circumscribed around the extreme positions of the manipulator slave arm structural elements are also plotted. The ratios of the manipulator service areas and volumes for each section to the corresponding values of areas and volumes of the circumscribed parallelepiped give us the values of the area service coefficients K_π and volume service coefficients K_o, which are put down in the table as well.

The value of the volume service coefficient corresponding to the complete utilization of the manipulator coverage determines the manipulator kinematic coefficient K_M.

The diagrams and plots of service areas and volumes so obtained facilitate the detailed qualitative and quantitative evaluation of the manipulator coverage and provide sufficiently objective information on their space performance.

Summary data on the space performance of some manipulator models are given in the table below. For the model "Compact" manipulator the possibility of its slave arm disagreement relative to the master arm, to enlarge the coverage is taken into account.

Table

Space performance of some manipulator models

Manipulator type	Model	v_M	v_o	V_M	V_o	V_c	k_M	K_M
				(m^3)				
Articulated tongs	МЛП МЦ-П-5	0,18 0,38	0,71 0,90	0,33 0,63	1,07 1,34	0,05 0,15	0,26 0,42	0,32 0,47
Telescopic manipulator	M22 Compact	1,55 2,16	3,65 4,84	2,18 3,36	5,05 6,46	0,78 0,96	0,42 1,45	0,43 0,52
Articulated manipulator	MM	0,91	1,99	1,38	2,56	0,45	0,46	0,54

4. Conclusions

The analysis of the foregoing leads us to the following conclusions:

1. Articulated tongs without a wrist unit have the least values of the coverage and the kinematic coefficient. The addition of a wrist unit can considerably enlarge the coverage.
These manipulators should be used in small working volumes (e.g. boxes).

2. Telescopic manipulators have large coverage, but they have large overall dimensions as well.

The kinematic coefficient values are $0,4 \div 0,5$. These manipulators should be used in large working volumes (in cells and large boxes).

3. Articulated manipulators have the best kinematic characteristics. The kinematic coefficients of this group manipulators are the highest ($0,5 \div 0,6$) while the service area is large enough. It is advisable to use these manipulators, when a large service area is required, but working volume dimensions are limited.

4. Additional adjustment motions of the slave arm can considerably enlarge the manipulator coverage (see data on "Compact" manipulator).

5. The utilization of the suggested method for studying the manipulator space performance permits us to design new master-slave manipulators with good engineering properties.

REFERENCES

[1] КОБРИНСКИЙ А.Е, СТЕПАНЕНКО Ю.А.- Некоторые проблемы теории манипуляторов, в Сб."Механика машин", вып.7-8, Изд-во "Наука", 1967,
стр.4-23.

[2] ЛАКОТА Н.А.,ЛОБАЧЕВ В.И.- Некоторые вопросы проектирования дистанционно управляемых копирующих манипуляторов, в Сб."Механика
машин", вып.27-28, Изд-во "Наука", 1971, стр.17-29.

[3] ЛИКИШОВ Г.И.- Анализ конструктивных схем копирующих манируляторов,
в Сб."Вопросы атомной науки и техники", серия "Проектирование",
вып.3, ЦНИИатоминформ, 1970, стр.33-38.

[4] Campbell I.C., Kosyakov V.N., Vertut J., Manual on Safety Aspects of the Design and
Equipment of Hot Laboratories, International Atomic Agency, Safety Series, No. 30,
Vienna .

[5] Vertut, J., La telemanipulation, "Industries Atomiques", 1965, No. 9-10, pp. 65-78.

[6] ВИНОГРАДОВ И.Б., КОБРИНСКИЙ А.Е., СТЕПАНЕНКО Ю.А.,ТЫВЕС Л.И.- Осебен-
ности кинематики манипуляторов и метод объемом, в Сб."Механика
машин", вып.27-28, Изд-во "Наука", 1971, стр.5-16.

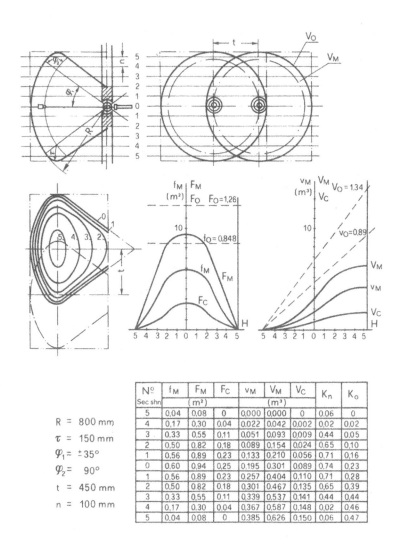

N⁰	f_M	F_M	F_C	v_M	V_M	V_C	K_n	K_o
Sec shn	(m²)			(m³)				
5	0,04	0,08	0	0,000	0,000	0	0,06	0
4	0,17	0,30	0,04	0,022	0,042	0,002	0,02	0,02
3	0,33	0,55	0,11	0,051	0,093	0,009	0,44	0,05
2	0,50	0,82	0,18	0,089	0,154	0,024	0,65	0,10
1	0,56	0,89	0,23	0,133	0,210	0,056	0,71	0,16
0	0,60	0,94	0,25	0,195	0,301	0,089	0,74	0,23
1	0,56	0,89	0,23	0,257	0,404	0,110	0,71	0,28
2	0,50	0,82	0,18	0,301	0,467	0,135	0,65	0,39
3	0,33	0,55	0,11	0,339	0,537	0,141	0,44	0,44
4	0,17	0,30	0,04	0,367	0,587	0,148	0,02	0,46
5	0,04	0,08	0	0,385	0,626	0,150	0,06	0,47

$R = 800\ mm$

$\tau = 150\ mm$

$\varphi_1 = \pm 35°$

$\varphi_2 = 90°$

$t = 450\ mm$

$n = 100\ mm$

Fig. 1 The study of the model MIII - Π - 5 articulated tongs coverage

THE SYNTHESIS AND SCALING
OF ADVANCED MANIPULATOR SYSTEMS

Carl R. FLATAU
Teleoperator Systems
Shoreham. N.Y. USA

(*)

Summary

The "Bilateral Force Reflecting" type of manipulator has been identified as the most dexterous of the presently available ones.

The extremely good man-machine integration inherent in these devices accounts for much of the superior dexterity. This superior performance does however make more stringent demands on most of the manipulator operating parameters. Of the many parameters involved, problems associated with size, weight, complexity and reflected inertia are traced through from a second generation device completed in 1969, to a new, third generation manipulator. The basic approach taken for the latter turns out to be quite suitable for sealing to smaller sizes. A design, scaled down by a factor of five, and intended for an "Artificial Intelligence" mini-robot arm as well as for a mini-manipulator, is described briefly.

(*) All figures quoted in the text are at the end of the lecture.

1. Introduction

The type of advanced manipulator under discussion here is variously described as a force reflecting master slave, or a bilateral force reflecting (BFR) manipulator. In the latter expression the term bilateral is used in the sense familiar to network synthesis specialists, and refers to the symmetric, bidirectional flow of power. For a BFR manipulator this means that in principle no distinction can be made between the master or Control Input Device (CID) and the slave or manipulator. A force vector input on either will result in a corresponding and proportional output vector of the other. The servo loops connecting the two are always so arranged as to retain this behavior. Indeed the above can be taken as a definition of the BFR principle.

The BFR manipulator is preferred since it has been shown that manipulators incorporating this principle are about an order of magnitude more dexterous than those incorporating the next best other control mode. [1, 2, 3] The explanation for the astonishing fact that the control mode alone can account for such a large improvement in dexterity is to be found in that a BFR manipulator probably constitutes the tightest man machine integration system invented so far. Where else could one find a system in which full proportional manual control in 28 parameters simultaneously can be accomplished by an uninitiated operator after only 1 minute learning time [2]. And after so short a learning time most operators perform at the 70 percentile effectiveness level. The 28 parameters obtain from a pair of BFR manipulators each having seven position parameters and seven force parameters.

It is becoming apparent from the above that more than kinematics is involved in the synthesis of a quality BFR servo manipulator. Indeed the disciplines involved are many. To mention just a few, one has human factors, man-machine integration, control theory, optimization procedures, kinematics, dynamics, energy management, machine design and more. Each of these must receive proper attention in manipulator synthesis, for the neglect of even one of these disciplines will surely result in a less than satisfactory manipulator. No attempt can be made here to cover the entire spectrum problems that must be dealt with in the synthesis of a BFR manipulator system   rather some specific topics will be highlighted and illustrated by means of several manipulator systems which have either been built by the author or are now in various stages of development.

The strong influence control mode on dexterity has been identified 2 . The role of other parameters is less clear. An attempt is therefore made to extract from each new manipulator design a maximum of new data on possible ways to furhter improve dexterity.

The Brookhaven manipulator

This manipulator has been described in the literature [4, 5, 6, 7, 8] and will be mentioned here only as a point of reference. It was developed primarily to satisfy the requirements of a number of proton accelereators like the Brookhaven "Alternating Gradient Synchrotron". These require a compact, truly remote manipulator to help in maintaining the more highly active portions of these machines.

The primary goal at the initiation of the program was not the achievement of an ultimate manipulator — this would take some decades anyway — but to obtain the performance levels of an "Argonne E-3" [9], or "Mascot" [10] in a much more compact package, so that it can be used in the narrow and low tunnels of typical accelerator installations. In point of fact, it was possible to achieve much improved performance in several respects as will be shown below.

The arrangement of the DOF (degrees of freedom) was changed but little from E-3 or Mascot. It is shown in Fig. 1. Only one DOF was changed primarily to ease counterbalancing.

The manipulator is shown in Fig. 2 with an operator at the CID (control input device) for size comparison. The compactness was achieved primarily by choosing permanent magnet, direct current, torque motors as actuators. In 1965 when these motors where chosen they were found to have the most nearly optimal characteristics of all motors available at the time. For the design pursued at that time, a combination of small size and weight coupled to a large output stall torque and a high power rate was desired. To improve the power rate even further, two motors were used in series in the manipulator while only one was required in the lesser capacity CID (also called master arm).

The importance of power rate can be illustrated by some very simple mathematics. If one calculates the effective mass (mo) at the tip of the manipulator due to motor rotor inertia Jm one gets:

$$m_o = Jm \left(\frac{N}{L} \right)^2 \tag{1}$$

where N is the gear ratio and L the moment arm from a given pivot to the tip of the manipulator.

Similarly the output torque "To" required to support a weight equal to the capacity "K" against gravity is given by:

(2) $$\text{To} = \text{K} \times \text{L} = \text{Tm} \times \text{N}$$

where Tm is the motor torque.
From (2) one gets:

(3) $$\frac{\text{N}}{\text{L}} = \frac{\text{K}}{\text{Tm}}$$

Which when substituted into (1) yields:

(4) $$m_o = K^2 \frac{Jm}{T_m^2}$$

Defining the well known parameter power rate "$\dot{P}$" as:

(5) $$\dot{p} = \frac{Tm^2}{Jm}$$

and substituting (5) into (4) one gets:

(6) $$m_f = \frac{K}{\dot{p}}$$

where m_f has been defined as $m_f = m_o/K$ or the motor inertia translated to the tip as an effective output mass and expressed as a fraction of capacity. The expression is also seen to be independent of gear ratio, a well known property of the parameter $\dot{p}$. Above expression must of course be adjusted to correspond to the system of units one wishes to use. Expression (6) is a very simple relationship which allows one to select motors of sufficiently low effective inertia.

As can be seen from Fig. 2, the 3 rotary DOF motors and gear reducers are distributed in the lower arm, the terminal device motor and gear reducer is in a small box which participates in the pitch (elevation) motion and is self counter-balancing. Two of the shoulder motion motors and reducers are on the upper arm behind its pivot point and are helping to counterbalance it. The last shoulder motor and reducer is mounted in a stationary box above the shoulder fork. Two counterweights, one for the upper, and one for the lower arm complete the arrangement. The 3 shoulder DOF use Harmonic drives (simple compact gear reducer using a flexible gear, developed by the united Shoe Machinery Corporation) for speed reducers. An exploded view of the shoulder motion is shown in Fig. 3. As can be seen from this illustration, there is a torque transducer at the output of each DOF at both the manipulator and the CID.

Due to the presence of the torque transducers it was possible to connect the servo loops according to the schematic of fig. 4. This schematic is of course highly simplified for clarity and shows only one DOF. This control scheme has the property of attenuating those reflected inertia components that are on the motor side of the torque transducer. Thus the total reflected inertia is about a factor or two less than that of previous BFR servo manipulators [9, 10] . This is so despite the rather large inertia due to the gross weight of the motors built into the lower arm

Another novel features not available in previous manipulators are:
1) Counterbalanced pitch and yaw (elevation and azimuth) motions.
2) Availability of a second mode for roll (twist) motion which allows continue rotation with reaction torque reflected back to the CID handle.
3) Logarithmic force reflection from manipulator to CID.
4) Extremely large motion excursions, as can be seen from comparison of Fig. 5 and Fig. 2 and also from the multiple exposure of Fig. 6.
5) About 1/3 the weight of previous BFR-servo manipulators.

Disadvantages of the Brookhaven manipulator

While the performance of the Brookhaven manipulator was equal or better than previous available devices, it was soon recognized that considerable improvements can be achieved. In particular, experiments with reflected inertia showed that this parameter has a rather large influence on dexterity. With the Brookhaven manipulator it is possible to increase the reflected inertia by reducing the gain in the force loop. An increase in reflected inertia of up to four times the optimum can be achieved. By running prescribed and well defined tasks with various inertia levels, and comparing a "Dexterity Quotient" $(DQ)^2$ as a figure of merit of dexterity, one found that in the case at hand a factor of four increase in reflected inertia would result in over a factor of 3 decrease in dexterity. One easily conjectures that a further decrease of reflected inertia might result in an improvement of dexterity. The correctness of this conjecture has now been verified by Vertut in his experimental work on dexterity [1,3]. It was concluded that even lower inertias than available with the Brookhaven manipulator are desirable.

Indeed inertia ratios "R_J" of about 20 are desirable where one defines:

(7)
$$R_J = \frac{m_{FL}}{m_{NL}}$$

Where m_{FL} is the effective manipulator output inertia with a mass of weight K at the tip, while m_{NL} is the unloaded inertia. The latter can not become zero in practical cases. It is to be noted that:

$$\text{as } m_{NL} \to 0 \; ; \; R_J \to m_f$$

So that calculation of m_f gives a first approximation of the final R_J.

Another difficulty inherent in the Brookhaven arm is total weight. It weighs some 55 kg (120 lbs) over half of which is due to the added counterweights. Therefore a search for a better counterweight method was indicated.

Finally the Brookhaven manipulator is rather intrically packaged and quite difficult and expensive to manufacture. A much simpler and less expensive design is desired.

The SM-209 – A third generation manipulator

The first two disadvantages of the Brookhaven arm suggest that one should try to remove the four motor-gear reducer units for the wrist motions and terminal devices out of the forearm and place them so that one can use their masses as counterbalance weight. This has the further advantage that the forearm as well as the upper arm will require rather smaller counterbalance moments. This would also make it possible to use larger motors which are more capable of satisfying the necessary motor parameters. The resulting manipulator is shown in Fig. 7, drawn from the manufacturing drawing in isometric form. As can be seen it has been possible to find a simple scheme, that places the four wrist and the TD motor-gear reducer units inside an ariticulated counterbalance mass and also use a simple tape drive that transmits these motions into the lower arm and wrist. The wrist used is a commercially available one, thus further reducing manufacturing difficulties.

It has also been possible to arrange the three shoulder drive motor housing in a tight cluster, which is stationary with respect to the manipulator support. Thus only their rotor inertia but not their gross mass enters into the reflected inertia picture.

The motors chosen are considerably larger than those of the Brookhaven manipulator and therefore have much higher torque resulting in much lower gear ratios.

All these innovations result in a manipulator which is much easier to build, weighs about 20 kg (44 lb) for a 10 kg handling capacity and has an inertia ratio above 10. This might be improved to about 20 with further electronic compensation.

The motors and associated electronics are a considerable improvement over previous techniques. Because of that, some prototype motors and electronics were furnished to Vertut at the French CEA for their MA-22 project (15 to 20 kg handling capacity), which bears some similarity to SM-209 [11]. Both these manipulators were developed independently and in parallel but experienced a certain amount of cross fertilization due to cooperation of the principle investigators.

SM 209 is envisioned as a first of a family of BFR-servo manipulator and will hopefully also serve as a basis and test bed for furhter advances in the art.

A mini-arm

The counterbalance system and tendon system used in SM-209 seem suited for other purposes. The artificial intelligence community likes to work with robot arms driven by their computer systems. Some of the workers in the field noted regretfully, that the computers they were using were able to shrink in size while getting ever so much more powerful, that such behavior was notably absent from the robot arms employed. Indeed it is hard to see why robot arms for artificial intelligence should not be conveniently small, portable and easily workable on a desk-top.

In an attempt to satisfy the need for a rather small arm, the scaling laws applicable to size reductions in the SM-209 design were studied. SM 209 seemed to be a particular suitable point of departure as the mini-robot arm required force sensing in six DOF and could benefit very much from a counterbalanced low inertia design.

Most of the applicable scaling laws were found to be favorable for the downscaling of the SM 209 design. The resulting mini-robot arm design proposal is shown in Fig. 8. As can be seen both the upper and lower arm are only 10 cm (4") long. This arm could be computer driven as well as manually controlled from the CID of SM 209. It might well prove to be a forerunner to manipulators used specifically for down scaled tasks.

Acknowledgment

The author would like to thank his colleagues and their secretarial staff at CERN and CEA-SACLAY. Without their very generous help it would have been impossible to complete this paper while traveling in Europe.

REFERENCES

[1]   J. Vertut et al. "Contribution to Analyze Manipulator Morphology Coverage and Dexterity (elswhere in these proceedings).

[2]   C.R. Flatau et al. "Some Preliminary Correlations between Control Modes of Manipulators and their Performance Indices" Proceedings of the first National Conference on Remotely Manned Systems, 189, 1971.

[3]   J. Vertut et al. "Contribution to Define a Dexterity Factor of Manipulators" Proceedings 21st Conference Remote Systems Technology, to be published Nov. 1973.

[4]   C.R. Flatau "Development of Servo Manipulators for high Energy Accelerator Requirements" Proceedings 13th Conference Remote Systems Technology. 29 (1965).

[5]   C.R. Flatau "Proposed Remote Handling Methods for a Modified AGS" IEEE Trans.Nucl. sci., NS-12, 3, 668 (1965).

[6]   C.R. Flatau "Compact Servo Master-Slave Manipulators with Optimized Communication Links", Proceedings 17th Conference Remote Systems Technology. 154. (1969).

[7]   V.J. Kovarik. "Notes on Force Reflecting Systems", AGSCD Technical Note No. 80. Brookhaven National Laboratory, Accelerator Dept., (1967).

[8]   C.R. Flatau. Advancements in Teleoperator Systems NASA-SP 5081, 16, (1970).

[9] R.C. Goertz et al. "The ANL Model 3 Master-Slave Electric Manipulator —
     Its Design and use in a cave", Proceedings 9th Conference Hot
     Laboratory Equipment, 121, (1961).

[10] L. Galbiati et al. "A Compact and Flexible Servo System for Master-Slave
     Electric Manipulators" Proceedings 12th Conference Remote Sys-
     tems Technology, 73, (1964).

[11] C.R. Flatau, J. Vertut et al. "MA-22 — A Compact Bilateral Servo
     Master-Slave Manipulator" Proceedings 20th Conference Remote
     Systems Technology, 296, 1972.

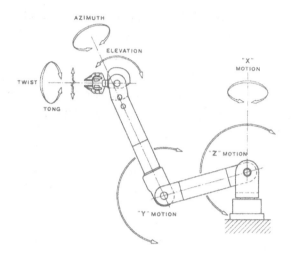

Fig. 1 - Arrangement of degrees of freedom
of the Brookhaven manipulator.

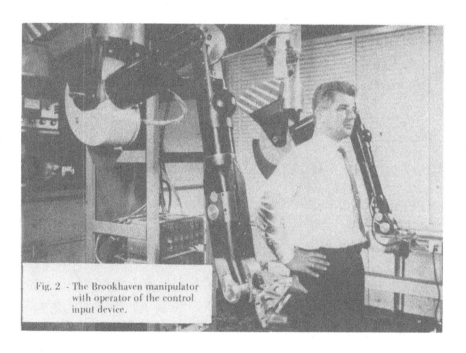

Fig. 2 - The Brookhaven manipulator
with operator of the control
input device.

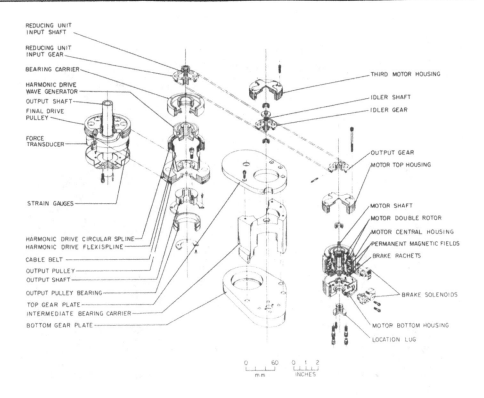

REDUCING UNIT
INPUT SHAFT

REDUCING UNIT
INPUT GEAR

BEARING CARRIER

HARMONIC DRIVE
WAVE GENERATOR

OUTPUT SHAFT

FINAL DRIVE
PULLEY

FORCE
TRANSDUCER

STRAIN GAUGES

HARMONIC DRIVE CIRCULAR SPLINE
HARMONIC DRIVE FLEXISPLINE

CABLE BELT

OUTPUT PULLEY

OUTPUT SHAFT

OUTPUT PULLEY BEARING

TOP GEAR PLATE

INTERMEDIATE BEARING CARRIER

BOTTOM GEAR PLATE

THIRD MOTOR HOUSING

IDLER SHAFT

IDLER GEAR

OUTPUT GEAR

MOTOR TOP HOUSING

MOTOR SHAFT

MOTOR DOUBLE ROTOR

MOTOR CENTRAL HOUSING

PERMANENT MAGNETIC FIELDS

BRAKE RACHETS

BRAKE SOLENOIDS

MOTOR BOTTOM HOUSING

LOCATION LUG

0   60   0 1 2
└─┴─┘ └┴┴┴┘
   m m   INCHES

Fig. 3   Exploded view of shoulder DOF Drives

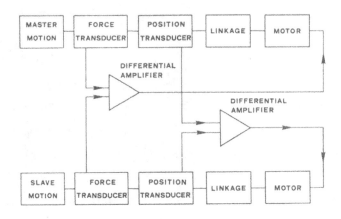

Fig. 4   Simplified control schematic

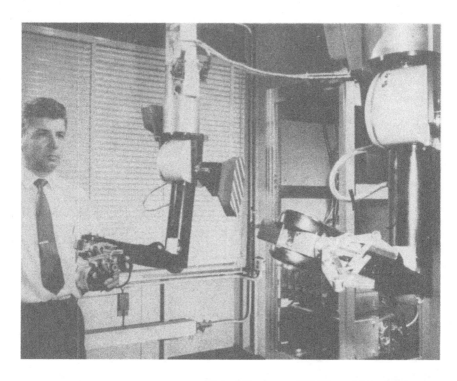

Fig. 5   An alternate manipulator stance

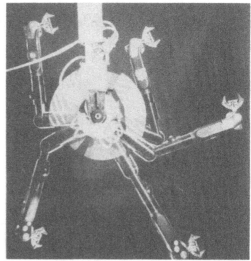

Fig. 6   Multiple exposure showing manipulator
motion range

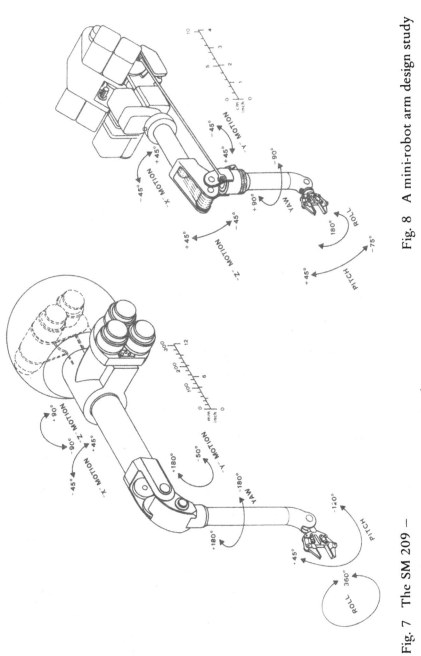

Fig. 8   A mini-robot arm design study

Fig. 7   The SM 209 –
A third generation compact manipulator

7. CONTROL OF MOTION

MASTER-SLAVE REMOTE-CONTROL MANIPULATOR

V.P. DOROKHOV, Department chief.
State Commettee for Utilization
of Atomic Energy, Moscow, USSR.

(*)

РЕЗЮМЕ

Предлагается принцип построения канала информации о нагрузке дистанционно управляемого копирующего манипулятора, исключающий влияние моментов неуравновешенных звеньев исполнительного механизма на точность воспроизведения усилий нагрузки на стороне оператора.Предлагается также введение в системы слежения дополнительных датчиков угла, расположенных вне управляющего механизма, например на пульте управления.Манипулятор, построенный с использованием этих предложений, позволит уменьшить физические нагрузки на оператора и повысить точность пространственной ориентации инструмента при выполнении станочных операций.

(*) All figures quoted in the text are at the end of the lecture.

Special remote-control power manipulators are required for performing different operations within dangerous areas thereby securing absolute safety for the operator.

For maximum efficiency of operator's work the general-purpose master-slave force-reflecting manipulators are most suitable, therefore in recent years in world manipulator development practice a trend has appeared to increase the fraction of force-reflecting manipulators in the scope of newly developed designs.

However, these manipulators are relatively complex and expensive, and at the same time the imperfection of designs, as well as the deficiencies of control and balancing systems cause rapid exhaustion of the operator.

Analysis of available manipulator operations shows that in most cases high efficiency of master-slave remote manipulators and convenience of work for the operator are possible without proportional power feedback.

The only requirement is to provide operating forces low enough not to cause object and slave arm damage.

For experimental research on this possibility we have developed and tested a master-slave electromechanical manipulator with force limited slave arm and limited movements of the master arm.

The manipulator control systems are as follows:
— slave follow-up systems;
— a load information system for the slave arm elements;
— a system of indirect unbalance compensation for the slave arm elements;
— a precise space orientation system for the slave arm elements;
— an instrument feed movement control system.

The load information channel design is based on exclusion of any action of intermediate load forces and on full locking of the corresponding elements of the master arm after the slave arm has reached the force which corresponds to the manipulator object rigidity. Such load information absolutely excludes any possibility of object rupture. The absence of intermediate force reproduction saves the operator from physical exhaustion in lengthy work.

Fig. 1 shows the block-diagram of the slave follow-up system for a single slave movement of the manipulator.

The operation principle of the system is as follows: an amplified summary signal of the angle transmitters 1 and 2, and of the speed transmitter 7 actuates the slave arm drive 9 and passes through electronic system model 10, whose transmission function approaches that of the slave arm during no-load operation.

A signal difference of the speed transmitter 7 and of the model 10, being proportional to the slave arm load and amplified by the device 11 drives the lock mechanism 6, which through the coupling 5 and the reduction gear 3 prevents the master unit from further movement. With the aid of the signal-level regulator 12 the operator sets the maximum permissible force of the slave arm.

An important feature of the manipulator is the load-information systems interconnection (Fig. 2), which permits unbalanced elements of the power and, hence, heavy-duty structure of the slave arm.

This is achieved by the introduction of cross load reflection channels, which interconnect the balanced elements of the slave arm (within adequate practical accuracy these are elements of the grip member) and all the master arm elements, which correspond to the unbalanced elements of the slave arm, through signal comparison blocks of balanced and unbalanced elements moment transmitters.

The master (1) and slave (2) arms are intercoupled through:
– control channels by position (3);
– direct channels of load reflection (4);
– cross channels of load reflections (5).

The direct load-reflection channels (4) include moment transmitters of the unbalanced (6) and balanced (7) elements of the slave arm as well as moment receivers (8) of the master arm elements, the transmitters (6) being tuned to maximum permissible forces of the corresponding slave arm elements.

The cross load reflection channels (5) are the connections between the moment transmitters (7) of the balanced elements and the moment receivers of the unbalanced elements achieved through the signal comparison blocks (9) of balanced and unbalanced elements moment transmitters.

With such arrangement the manipulator operates as follows: The master arm (1) elements' position in space is continuously followed up by the slave arm (2) elements' through the control channel by position (3). Loads within the slave arm elements are received by the moment transmitters 6 and 7, and through the direct load reflection channel (4) reach the master arm (1) moment receiver 8 in the element by elementary means. At the same time, forces within the balanced elements reach through the cross load reflection channels (5) and unbalanced element moment receivers (8) from the signal comparison blocks (9) of the moment transmitters of balanced and unbalanced elements. The maximum one of the signals being compared acts upon the moment receiver of each unbalanced element. This interaction of the load reflection channels prevents the moment influence of unbalanced elbow and shoulder movements of the manipulator on the quality of load reflection.

The precise space orientation system for the slave arm elements is realized through the duplicating angle transmitters 13 (fig. 1) of the power follow-up systems, which provide for independent control of each element from the control panel. For this to be done an instrument is fixed in the slave arm grip and is fed into the working area, the operator fixing the position with locks within the master arm joints and driving the instrument to the fixed point in space with the duplicating angle transmitters on the control panel. The slave-arm rigidity is achieved by energizing the brakes 14.

This control system principle is used as the basis of the remote-control power manipulator having the following characteristics:

Number of movements	– 8
Lifting capacity	– 50 kg
Torque	– 10 kgm
Service area radius	– 1.6 m

Instrument displacement with linearity maintaining:

horizontal direction	– 1.35 m
vertical direction	– 1.55 m
lifting capacity limit	
scale of the manipulator	– 10, 20, 30, 40, 50 kg

The manipulator is of the joint-lever kinematic design with drive motors, reduction gears, angle transmitters and brakes built into the slave arm joints and elements. It may be installed on different carriers (both horizontal and vertical).

Fig. 3 shows kinematic circuits of the master and slave arms.

The manipulator has seven feedback systems with limitation on forces, providing for the shoulder rotation (I) and tilting (II), the forearm tilting (III) and rotation (IV), the grip tilting (V) and rotation (VI) and the tongs movements (VII). There is an additional instrument feed movement, carried out from the control panel through an open control system with feed force measuring.

The master arm of the manipulator is a space joint multielement mechanism with independent kinematic circuits for all the movements, including angle transmitters and locks. Since with the load information principle described the operator has no need to expend much force for the master arm when manipulating, as any elements can move only to the lock point (with a preset force of the slave arm), great rigidity of the design is not required. Furthermore as the elements of the master arm joints have small size and low weight, and because of the possibility of reducing the size of the master arm members as compared with the slave arm, the master arm is of compact form and of low weight.

Thus, the power manipulator control principle and load information method suggested permit us to minimize the operator's physical forces and to develop different load-lifting capacity manipulators using a master unit of unified design and unified control system.

REFERENCES

[1]   КУЛЕШОВ В.С.,ЛАКОТА Н.А.- Динамика систем управления манипулятора-
       ми.Изд-во "Энергия", 1971.

[2]   БОР-РАМЕНСКИЙ А.Е.,КУЛЕШОВ В.С.,ЛАКОТА Н.А.,ЛОБАЧЕВ В.И.- Некоторые
       принципы построения дистанционно-управляемых копирующих манипу-
       ляторов.Сб."Механика машин", вып.7-8, Изд-во "Наука", 1967.

[3]   ЛАКОТА Н.А.,ЛОБАЧЕВ В.И.- Некоторые принципы проектирования дистан-
       ционно-управляемых копирующих манипуляторов.Сб."Механика машин",
       вып.27-28, Изд-во "Наука", 1971.

[4]   КОГАН-ВОЛЬМАН Г.И.- Передачи гибкими проволочными валами.Справочник.
       Машгиз, 1962.

[5]   ФРОЛОВ Л.Б.- Измерение крутящего момента, Изд-во "Энергия", 1967.

[6]   ПЕТРОВ Б.А.- Симметричная обратимая следящая система с моделями.Сб.
       "Применение инвариантных систем автоматического управления",
       Изд-во "Наука", 1970.

[7]   ДОРОХОВ В.П.,САДОВСКИЙ О.А, СМИРНОВ О.И.- Следящая система с отра-
       жением усилия.Авт.свид.№ 317039, Бюллетень изобретений, 1971,
       № 30.

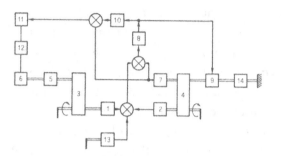

Fig. 1

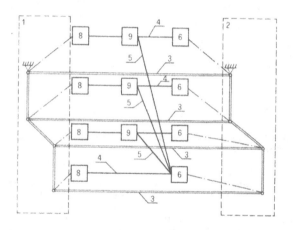

Fig. 2

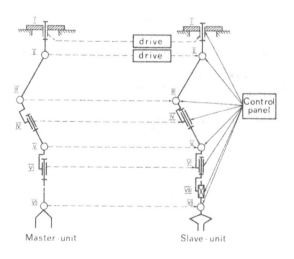

Fig. 3

NEW DEVELOPMENTS IN SYNERGIC RATE CONTROL OF MANIPULATORS[*]

M. GAVRILOVIĆ and M. MARIĆ
Mihailo Pupin Institute, Belgrade, Yugoslavia

(**)

Summary

In the development of manipulators the essential problem is the control synthesis. The concept of manipulator control depends to a great extent on its application. In cases when the neuromuscular complex of the human arm cannot be used as a source of control signals, the organisation of the control system becomes complex. For such manipulator applications the authors have proposed a concept of synergic rate control. This concept enables a synthesis of such a control system by which the operator can easily produce complex functional movements of the manipulator since he does it with a reduced number of control signals. In this paper new achievements in the development of this concept are presented. Special attention has been given to the construction of eating movements needed in rehabilitation. A computer-manipulator system was applied to evaluate ideas presented here.

(*) This work was supported by U.S. National Science Foundation Grant GF—31948.

(**) All figures quoted in the text are at the end of the lecture.

1. Introduction

The basic problems encountered in the development of rehabilitation manipulators are: design of the manipulator mechanism and its attachment to the operator (patient), selection of control sites and synthesis of the control system. However, the problem of control is the most essential one as well as the most difficult to solve.

Since the neuromuscular complex of the human arm of the operator cannot be used as a source of control signals, control of rehabilitation manipulators is inherently non-manual. The orthotic manipulator developed at the Rancho Los Amigos Hospital [1] at an early stage of development has indicated a significant lack of direct rate control. Namely, control of that manipulator requires too much conscious effort by the operator. In addition, the operator is absolutely unable to perform certain essential movements for which coordinated action of many control sites is required. For such manipulator applications the authors of this paper have proposed the concept of synergic rate control. According to this concept the operator controls selected generalized rate parameters by which the desired functional coordinated movements can be produced with a minimum of conscious effort. In the first paper which presented this concept [2], construction of two functional movements generated via the endpoint velocity vector of the manipulator is considered, one with tangential, the other with translational automatic coordination of the hand. The proposed concept gives the manipulator multifunctionality and partial autonomy, properties which are significant for efficient manipulation. Multifunctionality is reflected in the capability to perform several different types of functional movements, and autonomy in the fact that while a movement is being performed, a certain number of degrees of freedom of the manipulator are under automatic control.

The proposed concept was evaluated using a computer-manipulator system (Fig. 1) developed at the Pupin Institute. The control system developed for this eight-degree-of-freedom manipulator was realized as a two-level hierarchical structure. The level of regulators was realized as a set of eight electronic regulators for the angular positions of the manipulator joints, and the synergic level as a computer program. The system realization and the experimental results were presented in papers [3, 4]. In these experiments the operator used a joy-stick capable of producing one on/off signal and three continuous control signals.

In further development of the synergic rate control concept, rate control of the hand along the coordinate axes of the rectangular and spherical systems was included in the repertoire of movements. In this case the control system was so conceived that only one continuous and a single on/off control signal are required. With the digital signal the type of movement is selected, and with the analog signal the algebraic value is assigned to the selected rate parameter. In these experiments the operator carried out non-manual control of the manipulator by employing movements of his left and right shoulders. The system realization and the results obtained were given in paper [5]. In principle the synthesis of synergic rate control can be achieved by one of the following two-level hierarchical control structures.

In the first structure, the level of regulators consists of servos which control angular positions of the manipulator joints. The synergic level in this structure incorporates a mode selector, a set of control signal integrators, and a set of coordinate convertors.

In the second structure, the level of regulators consists of servos which control angular velocities of the manipulator joints. The synergic level in this structure incorporates a mode selector, and a set of coordinate convertors.

The concept of synergic control based on the second structure is briefly described in this paper in order to treat some control and computational problems. The problem of control synthesis with constraints on angular velocities of the joints is considered, and an efficient solution proposed for the first time. Furthermore, the construction of a new type of movement is introduced. This movement is functional in eating which is very important in rehabilitation.

Non-manual control of manipulators is studied in many research centers. The most important achievements in the development of control concepts, complementary to the one presented here, was reported by Whitney [6, 7]. Other important contributions were reported by Moe and Schwartz [8], and Potter and Freidman [9]. Singificant results in developing non-manual control of rehabilitation manipulators were obtained by Wirta and Taylor [10], and Simpson [11]. However, these systems were intended for particular cases when the operator is able to generate a sufficient number of naturally synergized control signals as the control systems for these manipulators do not incorporate artificial synergy.

2. Concept of Synergic Rate Control

The concept of synergic rate control that will briefly be presented here refers to the manipulator shown in Figure 1. The manipulator kinematics is symbolically depicted in Figure 2. The manipulator mechanism is an open chain made of levers l_1, l_2,...,l_7 connected by one-degree-of-freedom rotational joints J_1, J_2,...,J_7. The hand is attached to the manipulator tip. It consists of levers l_8' and l_8'' and a rotational joint J_8. The angular positions of the joints are denoted by θ_1, θ_2,..... θ_8 as indicated in Figure 2. The joints are actuated by d.c. electric motors placed on the manipulator support. Transmission of drive was realized with Bowden cables.

Direct non-manual control of the manipulator is very difficult because it requires too much conscious effort of the operator. The difficulties arise for the following reasons:

(1) Due to a limited capability for prediction the operator has troubles even controlling directly a mechanism with one degree of freedom (assuming that the information about the current position of the mechanism is received visually). The difficulties in control increase with the extent of irregularity of external forces which influence the mechanism motion, and the complexity of actuator transfer functions. The operator's capability for prediction is considerably increased if servodriving is applied. It has been proven that the operator positions the manipulator into a desired configuration most efficiently (with respect to time) when he controls its position, less efficiently when he controls its velocity, and least efficiently when he controls its acceleration.

(2) The operator has a limited capability for generating simultaneously several continuous control signals if naturally non-synergized muscle units are used. It becomes more and more difficult to train an operator to produce coordinated control signals as their number increases. However the operator could be trained to produce three such signals at the most. But if naturally synergized signal sources are used (e.g. the muscular complex of the arm), the operator's capability to generate control signals greatly increases. Unfortunately, in many cases only non-synergized muscle complexes are available.

(3) The operator receives visual information about the manipulator configuration in an integral form. Only by analysing the image can the operator extract the data about the angular positions of the joints. Such an analysis however, requires conscious effort of the operator and it is also time consuming.

It is now evident why direct control of the manipulator with many degrees of freedom (eight, for instance) is practically impossible when naturally non-synergized signal sources are used. In order to overcome the problem of multi-dimensionality, sequential control should be adopted to a certain extent. The extreme case is encountered when only one continuous signal is used to control each degree of freedom on a time-sharing basis. In a preferable case the operator has two or three naturally synergized signals available. For controlling the manipulator, coordinated activity of the synergized source complex can be shared among the manipulator subsystems (for example, to control the hand or the manipulator).

Non-synergized sources of signals are most efficiently used to specify the reference rates to the manipulator which are realized by appropriate rate servo-systems. However, switching the control from one to another rate parameter does not cause any change of the corresponding positional parameters.

The rate coordinates of the manipulator are defined as follows:

$$\omega_i = \frac{d\theta_i}{dt}, \quad i = 1, 2, \ldots, 8 \tag{1}$$

Therefore, the manipulator velocity can be described by the vector

$$\underline{\omega} = [\omega_1, \omega_2, \ldots, \omega_8]^T$$

The elementary movements which the operator can produce controlling directly individual rate coordinates ω_i have poor functional values. As far as coordinated movements are concerned the situation is even worse, certain functional coordinated movements are absolutely impossible for him to produce (e.g. the movement in which the hand is kept in a constant angular position in space).

However, by employing the concept of synergic rate control, the problem of producing functional movements can be solved efficiently even with a single control signal applied. How is such a significant reduction in the number of control signals achieved ? This question will be answered here by choosing as an illustration a set of functional movements (as fully formulated in Ref. 8) which can be easily realized through synergic rate control.

From the stand-point of functionality, the manipulator can be considered as consisting of two subsystems, the hand and the manipulator as a hand carrier. In order to determine the position of the hand in space the reference system of coordinates $(Oxyz)_R$ is attached to it as shown in Figure 3.

The motion of the hand in respect to the absolute system of coordinates $Oxyz$ is defined by the velocity vector $\underline{V}$ of the hand reference point O_R, and the rotation rate vector $\underline{\Omega}$ of the hand reference system $(Oxyz)_R$. The components of these vectors with respect to the hand reference system $(Oxyz)_R$ are of particular importance because of the functionality of movements they generate when specified. The movements along x_R, y_R, and z_R axes are produced by V_{x_R},

V_{y_R} and V_{z_R} respectively. The rotations about x_R, y_R, and z_R axes are generated by Ω_{x_R}, Ω_{y_R}, and Ω_{z_R} respectively. Each of these movements has a high degree of functionality in manipulation.

The specified velocity $\underline{V}_R$ and the rotation rate $\underline{\Omega}_R$ of the hand can be produced with a manipulator of six degrees of freedom. The manipulator shown in Figures 1 and 2 is redundant — it has seven degrees of freedom. However, this redundancy becomes functional if an appropriate parameter is placed under the operator's control, for example the angular velocity ω_1, $q = \omega_1$. With this control parameter the operator can place the elbow plane into the position which is the most suitable for manipulation. Placing prehension under the operator's direct control, $p = \omega_8$, is also functional.

The set of generalized rate coordinates whose values can be specified by the operator, forms the vector

(2) $$\underline{w} = [\, q, \underline{V}_R, \underline{\Omega}_R, p\,]^T$$

The vectors $\underline{w}$ and $\underline{\omega}$ are related by the following matrix equation

(3) $$\underline{w} = \underline{J}\,(\theta)\,\underline{\omega}$$

where $\underline{J}(\theta)$ the square matrix depends on the manipulator configuration. The elements of that matrix are composed of the partial derivatives:

(4) $$J_{ij} = \frac{\partial w_i}{\partial \omega_i}$$

Determination of vector $\underline{\omega}$, reduces to matrix inversion,

(5) $$\underline{\omega} = \underline{J}^{-1}\,(\theta)\underline{w}$$

This conversion of coordinates represents an artificial synergy of rate coordinates. Though the purpose of the conversion is evident, it is necessary to explain how with this or a similar synergy it is possible to reduce the number of control signals. The answer is very simple. The generation of the i-th elementary functional movement is performed by controlling the appropriate control velocity coordinate w_i. The values of the remaining seven coordinates are kept zero, $w_k = 0$, $k = 1,2,..., i-1, i+1,...,8$, and with respect to them no activity on behalf of the operator is required. For instance, by controlling the coordinate $w_2 = V_{x_R}$ (leaving the remaining components of the vector $\underline{w}$ at zero) the hand moves along axis x_R and during that movement its angular position in space is kept constant. By means of the component $w_3 = V_{y_R}$, and $w_4 = V_{z_R}$, the grasping device moves along y_R, and z_R axes, respectively. Through the component $w_5 = \Omega_{x_R}$ the grasping device is rotated about x_R axis keeping the reference point 0_R at constant position. Through components $w_6 = \Omega_{y_R}$, and $w_7 = \Omega_{z_R}$ the grasping device is rotated about y_R, and z_R axes, respectively. The component $w_1 = \omega_1$ rotates the manipulator about the axis $0x$ and during that time the position of the grasping device in space remains unchanged. Finally, with the component $w_8 = \omega_8$ the hand opens and closes.

The family of elementary synergized movements generated in this manner requires only one continuous control signal, and one discrete signal for movement selection. It is evident that the operator can readily compose complex manipulation movements from these eight elementary synergized movements as for example, reaching for a glass, bringing it to the mouth, drinking it up, and putting it back on the table. This complex movement cannot be performed by the operator if he controls directly the angular velocities of the manipulator.

When the operator has two naturally synergized sources of control signals available, the family of movements described is enriched with movements, performed by coordinated activity upon the selected pair of rate coordinates.

In principle, synthesis of the control described can be performed with or without feedback. Since the conditions in which the manipulator operates are variable, control systems without feedback are considered as unacceptable. Among various control systems with feedback, by means of which it is possible to realize manipulator synergic rate control, the simplest system to be synthesized is the one with an hierarchical two-level structure, consisting of the synergic and regulating control levels.

3. Synthesis of Synergic Rate Control

In order to simplify the synthesis, one coordinate system has been attached to each lever of the manipulator. The positions of these systems are defined for the initial manipulator configuration($\theta_i = 0$, $i = 1,2,...,8$) by translating the basic coordinate system $Oxyz$ along axis $0x$ into each manipulator joint (Fig. 4).

The vector description of the angular velocities ω_i in reference to $Oxyz$ are given by the following expression,

$$\underline{\omega}^i = \omega_i \, \hat{\omega}^i \,, \qquad i = 1,2,\ldots,8 \tag{6}$$

where $\hat{\omega}^i$ is a unit vector along the axes of the joint J_i.

Each unit vector $\hat{\omega}^i$ is expressed most simply in respect to its relevant coordinate systems $(Oxyz)_i$;

$$\hat{\omega}_i^i = \hat{x}_i \, , \qquad i = 1,7$$

(7)
$$\hat{\omega}_i^i = \hat{y}_i \, , \qquad i = 2,6$$

$$\hat{\omega}_i^i = \hat{z}_i \, , \qquad i = 3,4,5,8$$

with $\hat{x}_i$, $\hat{y}_i$, and $\hat{z}_i$ unit vectors of the coordinate system $(Oxyz)_i$.

The angular velocity of the hand $\underline{\Omega}_R$ is given by the following vector expression:

(8)
$$\underline{\Omega}_R = \sum_{i=1}^{7} \omega_i \, \hat{\omega}_R^i$$

The unit vector $\hat{\omega}_R^i$ is determined by mapping its original $\hat{\omega}_i^i$ from $(Oxyz)_i$ into $(Oxyz)_R$ system of coordinates. Since $(Oxyz)_R$ assumes the position of $(Oxyz)_7$ (see Fig. 4), the mapping expression takes the following form:

(9)
$$\hat{\underline{\omega}}_R^i = \underline{G}_i^7 \hat{\omega}_i^i \, , \qquad i = 1,2,\ldots,7$$

where $\underline{G}_i^7 = \underline{G}_i^{i+1} \underline{G}_{i+1}^{i+2} \ldots \underline{G}_6^7$ ($\underline{G}_i^{i+1}$ are well known rotational displacement matrices).

The velocity of the hand reference point O_R in respect to $(Oxyz)_R$ can be expressed as follows:

(10)
$$\underline{V}_R = \sum_{i=1}^{7} \omega_i \, \hat{\omega}_R^i \times \underline{r}_R^i$$

where $\underline{r}_R^i = (\overrightarrow{O_{i-1} \, O_R})_R$.
The vectors $\underline{r}_R^i$ are determined

(11)
$$\underline{r}_R^i = \sum_{k=i}^{7} 1_k \hat{1}_R^k \, , \qquad i = 1,2,\ldots,7$$

where 1_k denotes the length of k-th lever, and $\hat{1}_R^k$ describes its angular position in respect to $(Oxyz)_R$.

The unit vectors $\hat{1}^k$ are expressed most simply in respect to their relevant coordinate systems $(Oxyz)_k$,

$$\hat{1}_k^k = \hat{x}_k \quad , \qquad k = 1,2,\ldots,7 \tag{12}$$

The vectors $\hat{1}_R^k$ are determined by mapping,

$$\hat{1}_R^k = G_k^7 \hat{1}_k^k \quad , \qquad k = 1,2,\ldots,7 \tag{13}$$

According to Expressions 2,3,8, and 10 the following expression is obtained for the matrix $\underline{J}(\underline{\theta})$.

$$\underline{J}(\theta) = \begin{bmatrix} 1 & 0 & . & . & 0 & 0 \\ \hat{\omega}_R^1 \times \underline{r}_{-R}^1 & \hat{\omega}_R^2 \times \underline{r}_{-R}^2 & . & . & \omega_R^7 \times \underline{r}_{-R}^7 & 0 \\ \hat{\omega}_R^1 & \hat{\omega}_R^2 & . & . & \hat{\omega}_R^7 & 0 \\ 0 & 0 & . & . & 0 & 1 \end{bmatrix} \tag{14}$$

Inversion of matrix $\underline{J}(\theta)$ yields the desired transformation matrix $\underline{J}^{-1}(\theta)$.

The functional value of the described elementary movements comes from the fact that the operator easily estimates the position of the reference coordinate system $(Oxyz)_R$ because it is fixed to the hand in a characteristic way.

Another set of movements which are functional in manipulation is obtained when the velocity vector $\underline{V}$ and the rotation rate vector $\underline{\Omega}$ of the hand are specified with respect to the absolute coordinate system $Oxyz$.

Other sets of functional movements which can be realized by synergic rate control are described elsewhere [3,5,6,8,12] and will not be considered here. However, we will consider a new type of movement which has a significant functional value in eating activity (of importance in rehabilitation).

Mathematics encountered in synergic rate control of manipulators is treated in detail in works by Whitney [8] and Gavrilović [12].

4. Sheaf of Manipulator Movements

A new type of movement of great functional value is the movement along a trajectory leading through a prespecified manipulator configuration. A set of such trajectories forms a sheaf. The configuration through which pass all trajectories of the sheaf is called the focus. The functionality of this type of movement becomes eviden when it is necessary to bring the manipulator into a specific configuration from .various initial positions. Such a case is encountered with rehabilitation manipulator when food is brought to the mouth. The same type of movement is also necessary when objects, which are scattered in space, are to be collected by the manipulator and placed in the same position. A symbolic representation of such a complex movement consisting of several elementary movements of this type is given in Figure 5.

A convenient way to define the focus of the control system is to place the manipulator into the desired configuration $\underline{\theta}^*$ and to declare it as the focus. Denote an initial configuration of the manipulator by $\underline{\theta}^\circ$.

If ω represents a rate parameter specified by the operator, the expressions for the manipulator angular velocity is

$$(15) \qquad\qquad \underline{\omega} = \omega\,\hat{\omega}, \qquad i = 1, 2, \ldots, 7$$

where the unit vector $\hat{\omega}$ is determined by the focal and the initial configurations according to the expression

$$(16) \qquad\qquad \hat{\omega} = \frac{\theta^* - \theta^\circ}{|\theta^* - \theta^\circ|}$$

By defining the parameter ω the operator can perform the movement with desired speed, interrupt it at any moment, or generate a movement of inverse direction by changing the sign of parameter ω.

The sheaf of trajectories that can be obtained in this manner is a sheaf of straight lines in the multi-dimensional space of variables $\theta_1, \theta_2, \ldots, \theta_7$.

It is relatively simple, using the same concept, to synthesize a sheaf of such trajectories appearing as a sheaf of straight lines in another space.

5. Computational Aspects of Synergic Control

This section considers the most important computational problems appearing when performing synergic control.

Inversion of the Transformation Matrix $\underline{J}(\theta)$

Inversion of matrix $\underline{J}(\theta)$ requires particular attention because it is necessary to perform this inversion in real time as quickly as possible. For the manipulator mechanism shown in Figures 1 and 2, the matrix inversion can be simplified. Since the joints J_5, J_6 blend into one complex joint, the length of lever l_5 is zero. Owing to the fact that the axis of joint J_7 is placed along the axis $O_5 x_5$, that is $O_6 x_6$, joint $J_{\bar{7}}$ can be also considered as part of the same complex joint in which case the lever length l_6 is zero. If, in addition to this, the reference point O_R is placed into joint J_7, the reference system $(Oxyz)_R$ assumes the position of the system $(Oxyz)_7$. If follows that $|\underline{r}^5|, |\underline{r}^6|, |\underline{r}^7| = 0$. The matrix inversion can now be performed gradually.

ω^1 is specified directly,

$$\omega_1 = q_1 \tag{17}$$

Equation 10 reduces in this case to:

$$[\hat{\omega}_R^2 \times \underline{r}_R^2 \quad \hat{\omega}_R^3 \times \underline{r}_R^3 \quad \hat{\omega}_R^4 \times \underline{r}_R^4] \begin{bmatrix} \omega_2 \\ \omega_3 \\ \omega_4 \end{bmatrix} = \underline{V}_R - \omega_1 \hat{\omega}_R^1 \times \underline{r}_R^1 \tag{18}$$

from which $\omega_2, \omega_3, \omega_4$ are determined.

Equation 8 can be written as

$$[\hat{\omega}_R^5 \quad \hat{\omega}_R^6 \quad \hat{\omega}_R^7] \begin{bmatrix} \omega_5 \\ \omega_6 \\ \omega_7 \end{bmatrix} = \underline{\Omega}_R - \sum_{i=1}^{4} \omega_i \hat{\omega}_R^i \tag{19}$$

from which $\omega_5, \omega_6, \omega_7$ are computed.

ω_8 is specified directly,

$$\omega_8 = q \tag{20}$$

It is evident that the procedure for inverting matrix $\underline{J}(\theta)$ has been simplified and therefore speeded up.

When solving system 18 the question of singularity of matrix $[\hat{\omega}_R^2 \times \underline{r}_R^2 \quad \hat{\omega}_R^3 \times \underline{r}_R^3 \quad \hat{\omega}_R^4 \times \underline{r}_R^4]$ is raised. In case that $l_3 = l_4$ and $|\theta_4| < \Pi$ (Fig. 1), we have $|\underline{r}_R^i| > 0$, $i = 2,3,4$. Vectors $\hat{\omega}_R^2$ and $\underline{r}_R^i$, $i = 2,3,4$ lie in the elbow plane, whereas $\hat{\omega}_R^3$ and $\hat{\omega}_R^4$ are perpendicular to it. It follows that $\hat{\omega}_R^3 \times \underline{r}_R^3$ and $\hat{\omega}_R^4 \times \underline{r}_R^4$ also lie in that plane. Singularity occurs when they are colinear. Since $\hat{\omega}_R^3$ and $\hat{\omega}_R^4$ are always colinear, the condition reduces to colinearity of $\underline{r}_R^3$ and $\underline{r}_R^4$ which results when $\theta_4 = 0$. Another singularity appears when $\underline{r}_R^2$ is colinear with $\hat{\omega}_R^2$ since $|\hat{\omega}_R^2 \times v_R^2| = 0$. This occurs when $\theta_3 + \theta_4 = \pm \Pi/2$. Hence, the matrix is singular when

(21) $$\theta_4 = 0 \quad \text{and/or} \quad \theta_3 + \theta_4 = \pm \frac{\Pi}{2}$$

In solving system 19 the question of singularity of matrix $[\hat{\omega}^5_R \hat{\omega}^6_R \hat{\omega}^7_R]$ is raised. Since ω^5_R, ω^6_R and ω^7_R are unit vectors, the matrix can be singular only if some of the vector pairs are colinear. Since only colinearity of vectors ω^5_R and ω^7_R is possible, singularity occurs when

(22) $$\theta_6 = \pm \frac{\Pi}{2}$$

Presence of Constraints on Angular Velocities

Due to the limited capabilities of the driving motors only those angular velocities can be achieved, which satisfy the constraints:

(23) $$|\omega_i| < \omega_{i\ max}, \qquad i = 1, 2, \ldots, 7$$

If even one angular velocity exceeds the prescribed boundary, the movement performed by the manipulator would degenerate. This can be successfully avoided by a proportional reduction of all angular velocities so that none of the boundaries (Expression 23) are violated. This is achieved in the following manner. The maximum absolute value among relative angular velocities is determined,

(24) $$\lambda = \max \left(\frac{|\omega_1|}{\omega_{1\,max}}, \frac{|\omega_2|}{\omega_{2\,max}}, \ldots, \frac{|\omega_7|}{\omega_{7\,max}} \right)$$

If $\lambda > 1$, then the angular velocities are constrained to the following values:

(25) $$\omega_i^* = \frac{1}{\lambda} \omega_i, \qquad i = 1, 2, \ldots, 7$$

Presence of Constraints on Angular Positions

Due to design constraints, the angular positions of the manipulator joints are bounded to the values

(26) $$\theta_{i\ min} \leq \theta \leq \theta_{i\ max} \qquad i = 1, 2, \ldots, 8$$

If during a movement any of the manipulator joints violates the constraint, the movement degenerates. To prevent this, the angular velocities could be reduced to zero which causes the manipulator to stop moving. However, experiments have proved that it is more functional for manipulation to permit degenerated movements to occur.

6. System Realization

The concept of synergic rate control was evaluated using a computer-manipulator system developed at the Pupin Institute (Fig. 1). The entire complex consisting of a manipulator, a control system and a human operator is presented by a block diagram given in Figure 6. The control system was realized as a two-level hierarchical structure. The regulating control level was realized as a set of eight electronic regulators while the synergic control level was implemented by a mini-computer (Varian 620-i). The computer program was written in the assembly language to speed up computation (for other steps see Ref. [12]).

Three dimensional computer simulation of the manipulator in real time is also used as an alternative method in concept development and evaluation [13].

7. Conclusion

Synergic rate control of manipulator has been presented here, with particular attention devoted to the assence of the control concept itself. It has been shown how that control is realized by synergizing the rate coordinates of the manipulator. From the set of elementary functional movements whose synthesis was carried out, a subset of the most characteristic movements was selected in order to illustrate the concept. In addition, a new type of elementary movement has been described which is functional when it is necessary to transfer the manipulator from various initial configuration into one which is predefined. Such a case is encountered when food is brought to the mouth with a rehabilitaiton manipulator, or when the manipulator collects a set of scattered objects. Singularity of the transformation matrix has been analysed for the particular manipulator considered. The presence of constraints upon the angular velocities has been considered for the first time. It was shown how this problem can be solved efficiently.

Synergic rate control of manipulators is highly functional particularly when the operator has a minimum number of control signals available. However, it is interesting to consider synergic positional control for a manipulator, and also the combined synergic positional-rate control that have not been dealt with so far, or their functionality evaluated. Further development of the manipulator synergic control concept is of a broader significance because that concept in principle can be applied for solving the general problem of man-machine complexes.

REFERENCES

[1]   Allen, J.R., Karchak, A., Jr., Nichel, V.L., Snelson, R., "The Rancho Electric Arm", The Third Annual Rocky Mountain Bioengineering Symposium IEEE Conference Record F-62, 79-81, May 1966.

[2]   Gavrilović, M.M., and Marić, M.R., "An Approach to the Organization of the Artificial Arm Control", Proc. of the Third International Symposium on External Control of Human Extremities, ETAN, Belgrade, 1970.

[3]   Tomović, R., Gavrilović, M. and Marić, M., "Computer Coordinated Remote Manipulation", Proc. of the IFIP Congress 71, Ljubliana, North-Holland Publishing Company, 1972.

[4]   Marić, M.R., Gavrilović, M.M., and Radovanović, D., "Synergic Control of Computer-Manipulators", Proc. of the Fourth Symposium on Automatic Control in Space, ETAN, Belgrade, 1971.

[5]   Marić, M.R., and Gavrilović, M.M., "Evaluation of the Synergic Control of the Rehabilitation Manipulator", The Fourth International Symposium on External Control of Human Extremities, Dubrovnik, 1972.

[6]   Whitney, D.E., "Resolved Motion Rate Control of Manipulators and Human Prostheses", IEEE Trans. on Man-Machine Systems, Vol. MMS-10, No. 2, June 1969.

[7]   Whitney, D.E., "The Mathematics of Coordinated Control of Prostheses and Manipulators", The Fourth International Symposium on External Control of Human Extremities, Dubrovnik, 1972.

[8]   Moe, M.M., and Schwartz, J.T., "A Coordinated Proportional Motion Controller for an Upper Extremity Orthotic Device", Proc. of the Third International Symposium on External Control of Human Extremities, ETAN, Belgrade, 1970.

[9]   Potter, A.G., and Friedman, J.H., "End Point Control Using Hand Orientation", The Fourth International Symposium on External Control of Human Extremities, Dubrovnik, 1972.

[10] Wirta, R.W., and Taylor, R., Jr., "Development of a Multiple-Axis Myolectrically Controlled Prosthetic Arm", Proc. of the Third International Symposium on External Control of Human Extremities, ETAN, Belgrade, 1970.

[11] Simpson, D.C., "An Externally Powered Prosthesis for the Complete Arm", Bio-Medical Engineering, Vol. BME-16, March 1969.

[12] Gavrilović, M.M., "Hierarchical Multi-level Control of Manipulation and Locomotion Mechanisms", Ph. D. Thesis, Dept. of Electrical Engineering, University of Belgrade, Jan. 1973.

[13] Gavrilović, M.M., and Selić, B.V., "Three Dimensional Graphical Manipulator Simulation in Real Time", The Seventeenth Yugoslav Annual Meeting ETAN, Novi Sad, June 1973.

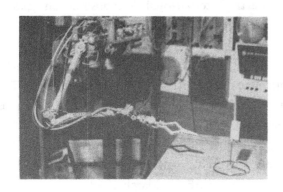

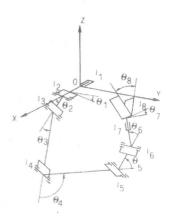

Fig. 1   Eight-degree of freedom experimental
manipulator built at the Pupin Institute

Fig. 2   Schematic representation of the manipulator kinematics

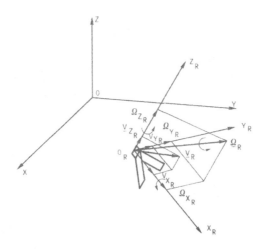

Fig. 3

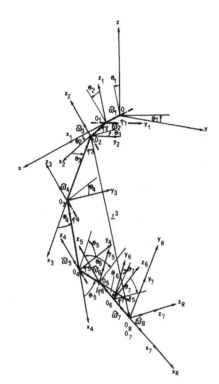

Fig. 4

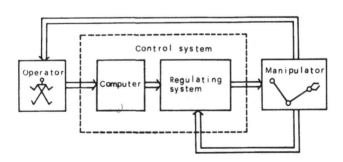

Fig. 5 Sheaf of manipulator movements

Fig. 6 Operator-control **system-manipulator** block diagram

ROBOT AND MANIPULATOR SLAVE FROM CONTROL VIEWPOINT

V.S. KULESHOV, Senior research worker of the
Academy of Sciences, USSR, Moscow, USSR.

V.N. SHVEDOV, Senior research worker of the
Academy of Sciences, USSR, Moscow, USSR.

(*)

Summary

In the paper the questions of robots' and manipulators' slave unit dynamics as control objects are under consideration. With allowance for robots' and manipulators' performance characteristics, the authors succeeded in getting the analytical and structural representation of motion transduced from an actuator to an appropriate joint axis.

Using the Lagrangian equation of the second-kind equations of robots' and manipulators' slave units dynamics as control objects are investigated. Such representation makes it possible to analise robots and manipulators as multidegree control system and to formulate design requirements.

(*) All figures quoted in the text are at the end of the lecture.

Robot and manipulator slave units have various motion transducers such as planetery and non-planetery gears, rack-and-gear drives, wave drives, ball-and-screw drives, rope, ribbon, chain, flexible and tubullar shafts, etc. They may be used as mechanical transmissions connecting slave actuator and dynamic link which serves as the load ("shoulder", "forearm", etc.). The selection depends on the type of slave actuators and their location on the slave, the kind of the hazardous area, the application and service life of the manipulator, etc. [17].

In many robots and manipulators, degrees of freedom are controlled by servo-drives which are packaged in one motor (or control) unit. The transmission of motion from an actuator to an appropriate joint involves different kinematic links. In task performance many degrees of freedom should be controlled simultaneously. With master and slave units restricted in external dimension, the mechanical transmission must be very compact. For this reason great attention is currently payed to hinge gears, gears with reversible power flexible wire shafts combined with ball-and-screw, [1]. In the latter the rotation of hinge gears or flexible shafts is transformed into the motion of levers changing the relative position of manipulation or control unit links.

The main characteristics of flexible wire gears are stiffness in flexure, torsional rigidity, and a factor allowing the taking into consideration of loss of the flexible shaft elasticity due to internal frictions under elastic deformation.

The elasticity of the mechanical transmission protects it against instantaneous overloads which are likely to occur in manipulator slaves (e.g., those caused by shocks). The elasticity in the control system load circuit is, however, found to deteriorate the operating dynamics and the system accuracy, [1].

The wire gear torsional rigidity $K_{e.f.}$ is found as the ratio of the transmitted torque T to elastic deformation (in the static state) γ

(1) $$K_{e.f.} = \frac{T}{\gamma}$$

Flexible wire geat elastic deformations tend to increase greatly in operational use and are thus time depedent.

One important parameter of such gears is a rotation allowing for loss in elasticity (increased rigidity) of the flexible shaft owing to internal frictions between wire turns and layers in elastic deformation. Indeed, the wire gear rigidity during elastic deformation, or in dynamic conditions, is found to be somewhat greater than in static tests. The increased rigidity can be approximately regarded proportional to the rate at which the relative position of winds and layers changes. Consequently, in dynamic conditions the flexible shaft elastic deformation will be given as

$$\gamma = \frac{T}{K_{e.f.} + \chi p} \qquad (2)$$

Studies revealed that parameters of an elastic gear are given by the ratio χ/K_{ef} = const. which is independent of the shaft length and is sensitive only to the flexible wire gear material.

The ball-and-screw converters in remotely controlled manipulators seem to be most promising elements as they eliminate gear transmissions or reduce the number of their pairs so as to reduce drastically the size of manipulators. The specific property of mechanical gears with ball-and-srew converters and flexible shafts is that in certain motions the gear ratio varies as a function of the drive positions in a range as wide as 500 through 1500.

Mechanical gears of robot and manipulator slaves may also include torque sensors which are needed in control systems using force feedbacks. The block-diagram of one such transmission is given in Fig. 1a with the following notations :

J, ns^2	— the engine rotor and kinematic transmission moment of inertia;
T_a , nm	— the torque developed by the actuator;
J_e = var, ns^2	— the load moment of inertia ;
α_a ; α_e , rad	— the engine and load shaft angle;
γ_1 ; γ_2 , rad	— elastic deformations in the kinematic transmission and the torque sensor, respectively;
$K_{e.f.}$; C, $\frac{nm}{rad}$	— the kinematic transmission and torque sensor elasticity factor;
χ, $\frac{ns}{rad}$	— a factor allowing for losses due to deformation;
2σ , rad	— the mechanical transmission play;
i_r	— the reduction gear ratio;
T_d , nm	— the external load torque;
T_f' , nm	— the torque of resistance to the engine shaft motion due to friction in seals, bearings, commutator brushes, etc.;
T_f'' , nm	— the torque of resistance to the motion of the load shaft due to dry and viscous friction;
T_{df} , nm	— the static torque caused by the elevated load and the weight of the slave kinematic links (depends nonlinearly on the turn angles of each degree of freedom).

As has been shown in a number of papers [1, 2] the description of the behaviour of a mechanical system allowing for its elastic properties and cleavances can be easily obtained. Forces or moments transmitted from one concentrated mass to another are generally a nonlinear function of elastic deformations and cleavances separating these masses.

This function can be represented in the following form

$$(3) \qquad T_e(t) = (K_{e.f.} + Xp)\gamma_1(t) = C \cdot \gamma_2(t)$$

The value of the elastic deformation of the overall mechanical transmission is given by the expression

$$(4) \qquad \gamma(t) = \gamma_1(t) + \gamma_2(t) = T_e(t) \cdot \frac{d(K_{e.f.} + Xp) + C}{d(K_{e.f.} + Xp) \cdot C}$$

Hence

$$(5) \qquad T_e(t) = \frac{d(K_{e.f.} + Xp) \cdot C}{d(K_{e.f.} + Xp) + C} \cdot \gamma(t)$$

Here γ is a nonlinear function allowing for the play and deformation and described by the obvious equation

$$(6) \qquad \gamma = \begin{cases} \dfrac{\alpha_a}{i_r} - \alpha_e - \sigma & \text{at } \dfrac{\alpha_a}{i_r} - \alpha_e > \sigma \\[2mm] 0 & \text{at } \left| \dfrac{\alpha_a}{i_r} - \alpha_e \right| \leqslant \sigma \\[2mm] \dfrac{\alpha_a}{i_r} - \alpha_e + \sigma & \text{at } \dfrac{\alpha_a}{i_r} - \alpha_e < -\sigma \end{cases}$$

With an allowance for eq. (5) and the torques α_e at the shaft, eq. (3) takes the form

$$(7) \qquad \begin{aligned} &J_e p^2 \alpha_e(t) + |T''_{d.f.}| \operatorname{sign} p\alpha_e(t) + f'' p\alpha_e(t) \pm T_d(t) + T_{d.f.}(\alpha_e) = \\ &= \gamma(t) \frac{d(K_{e.f.} + Xp) \quad C}{d(K_{e.f.} + Xp) + C} \end{aligned}$$

Ball-and-screw drives acting as force reducing gear make the gear ratio i_r vary depending on the load shaft turn angle. Figure 1b represents a kinematic diagram of one of the degrees of freedom where a ball-and-screw drive is used.

The gear ratio of such a transmission is given as follows :

$$i_r = \frac{T_e}{T'_e} \tag{8}$$

where T_e is the torque at the load shaft; T'_e is a torque reduced to the engine shaft

$$T_e = \frac{P}{K_{b.s.d}} \tag{9}$$

$$P'_e = \frac{T_e}{b} \tag{10}$$

$$P_e = \frac{T_e}{b} \cos \gamma \tag{11}$$

Here $K_{b.s.d.}$ is the gear ratio of the ball-and-screw drive, b and γ are as shown in Fig. 1b. $\cos \gamma$ is the load shaft turn angle function. From Fig. 1b it follows that

$$\left.\begin{array}{l} h = b \cdot \cos \gamma ; \\[4pt] h = a \cdot \sin \beta ; \\[4pt] \dfrac{b}{\sin \beta} = \dfrac{c + x}{\sin \alpha_j} \end{array}\right\}$$

$$\sin \beta = \frac{b \cdot \sin \alpha_j}{c + x} \quad ; \quad h = \frac{ab \cdot \sin \alpha_j}{c + x} \tag{12}$$

Hence

$$b \cos \gamma = \frac{ab \cdot \sin \alpha_j}{c + x} \quad ; \quad \cos \gamma = \frac{a \cdot \sin \alpha_i}{c + x} \tag{13}$$

Substituting eq. (9) into (8) and allowing for expressions (11), (12), (13) we have :

$$i_r = \frac{T_e}{T'_e} = \frac{K_{b.s.d.} \; b(c + x)}{a \cdot \sin \alpha_j} \tag{14}$$

We determine the displacement of the ball-and-screw drive rod when the engine shaft turns as

$$X = \frac{\alpha_a}{K_{b.s.d.}} \tag{15}$$

From a triangle with the sides a, b, c + x we find that

(16)
$$X_{1,2} = -(a-b) \pm \sqrt{a^2 - 2ab \cos\alpha_j + b^2}$$

Substituting eq. (16) into (14) and taking into consideration that $c = a + b$

(17)
$$i_r = \frac{K_{b.s.d.} \cdot b \cdot \sqrt{a^2 + b^2 - 2ab \cos\alpha_j}}{a \cdot \sin\alpha_j}$$

We introduce the notation

$$K_1 = \frac{K_{b.s.d.} \cdot b \cdot \sqrt{2ab}}{a} \quad ;$$

$$K_2 = \frac{a^2 + b^2}{2ab} \quad ;$$

Then

(18)
$$i_r = \frac{K_1 \sqrt{K_2 - \cos\alpha_j}}{\sin\alpha_j}$$

As a rule, in the zone of working angles the following conditions are met

$$\alpha_j \neq 0 , \quad \cos\alpha_j \ll K_2 .$$

Then

(19)
$$i_r = \frac{K^*}{\cos\alpha_e} \quad ;$$

where $K^* = K_1 \sqrt{K_2}$; $\alpha_e = 90° - \alpha_e$

Differential equation (7) with an allowance for eq. (19) corresponds to the diagram of Fig. 1c. This representation of slave mechanical transmissions is necessary for studies of autonomous systems for robot and manipulator control.

This, however is not sufficient for studies of the dynamics of manipulator and a robot controlled motion. The problem is that their slaves are multi-link mechanisms with an open kinematic circuit and complex interactions of joints in overall motion [3].

Depending on the kinematic relations in the joints, the slaves can be classified into two basic types : those with independent joint action (KIJA), Fig. 2a, b, and those with kinematically dependent joint action (KDJA), Fig. 3a, b.

The former type will be understood as slaves where motion along each degree of freedom does not cause corresponding motions in other degrees of freedom. This is the case of all manipulators with actuators situated directly in the hinges of the slave. If the actuators are in the motor unit, then motions to each joint are transmitted independently.

The latter type includes slaves where the motion of just one degree of freedom may cause related motion in others. This is the case of manipulators where actuators are placed in the motor unit, and of certain mechanical manipulators.

Let us consider analytical and structural representation of manipulator slaves as plants with an allowance for interaction of the degrees of freedom.

Analysis of the kinematic diagrams of slaves and of operations performed by the manipulator reveals that for the study of a control system with interaction it is sufficient to consider the motion of three basic degrees of freedom : the swing of the shoulder, elbow and wrist links in a plane. The vertical or its equilibrium, position of the slave is assumed to be the initial position.

In the light of the above, the kinematic diagram of a three-degree (KIJA) slave can be visualized as in Fig. 2c with the following notations :

α_e — the load shaft angle of each degree of freedom;

1_j ; m_j — the link length and mass;

τ_j — the distance from the link center of gravity to its axis of rotation;

T_{ej} — the load motion torque;

$j = 1, 2, 3...$ — the number of degrees of freedom.

The link mass includes the mass of the construction and the mass of actuator elements in the slave joints. The centres of gravity are assumed to stay on the longitudinal axes of kinematic elements. The constraints imposed on the system are assumed geometrical, or ideal.

The dynamics equation for such a system can be found as (of the second-kind) Lagrangian equations [4] generally given as

$$\frac{d}{dt}\left(\frac{\partial L}{\partial \dot{q}_j}\right) - \frac{\partial L}{\partial \dot{q}_j} = Q_j \qquad (20)$$

where

L $= K - \pi$ — the Lagrangian function,

K ; π — the system kinematic and potential energy,

Q_j — generalized torques of nonconservative forces,

q_j — a generalized coordinate.

For the mechanism under study

(21) $Q_i = -f'' p \alpha_{ej}(t) - |T''_{d.f.}| \operatorname{sign} p \alpha_{ej}(t) \pm T_{dj} + T_{ej}$

We must define the values of T and π incorporated into the second-kind equations. The system generalized coordinates will be the angles $\alpha_{\ell 1}, \alpha_{\ell 2}, \alpha_{\ell 3}$ between longitudinal axes of the corresponding links in a fixed system of coordinates X Y (Fig. 2c). The coordinates of the centers of gravity are denoted as $x_1 y_1, x_2 y_2, x_3 y_3$, respectively. Then the expression for kinematic energy is given as

(22) $K = \sum_{j=1}^{3} \frac{m_j v_j^2}{2}$

where $v_j^2 = \dot{x}_j^2 + \dot{y}_j^2$ is the squared velocity of the center of gravity of the j-th link. The potential energy is given as

$$\pi = \tau_1 m_1 g(1 - \cos\alpha_{\ell_1}) + m_2 g\{\ell_1(1 - \cos\alpha_{\ell_1}) + \tau_2[1 - \cos(\alpha_{\ell_1} + \alpha_{\ell_2})]\} +$$

$$+ m_3 g\{\ell_1(1 - \cos\alpha_{\ell_1}) + \ell_2[1 - \cos(\alpha_{\ell_1} + \alpha_{\ell_2})] + \ell_3[1 - \cos(\alpha_{\ell_1} + \alpha_{\ell_2} + \alpha_{\ell_3})]\}$$

(23)

where g is the free-fall acceleration.

To simplify the equations, consider linearized representation of the slave dynamics. At small deviations and under zero initial conditions one can assume that

$$\sin\alpha_{\ell j} \cong \alpha_{\ell j} \ ; \ \cos\alpha_{\ell j} \cong 1.$$

Substituting eqs. (21), (22) and (23) into L and solving eq. (20) for the generalized coordinates we will have for a system of three-degree actuators on robots and manipulators slave units with KIJA

$$a_{11} p^2 \alpha_{\ell_1} + a_{12} p^2 \alpha_{\ell_2} + a_{13} p^2 \alpha_{\ell_3} + c_{11} \alpha_{\ell_1} + c_{12} \alpha_{\ell_2} + c_{13} \alpha_{\ell_3} +$$

$$+ f_1'' p \alpha_{\ell_1} + |T''_{d.f.}| \operatorname{sign} p \alpha_{\ell_1} \pm T_{d_1} = T_{\ell_1} ;$$

$$a_{21}\, p^2 \alpha_{\ell_1} + a_{22}\, p^2 \alpha_{\ell_2} + a_{23}\, p^2 \alpha_{\ell_3} + c_{21}\, \alpha_{\ell_1} + c_{22}\, \alpha_{\ell_2} + c_{23}\, \alpha_{\ell_3} +$$

$$+\, f_2''\, p\, \alpha_{\ell_2} + \left| T_{d.f.}'' \right|\, \text{sign } p\ \alpha_{\ell_2} \pm T_{d_2} = T_{\ell_2}\,;$$

$$a_{31}\, p^2 \alpha_{\ell_1} + a_{32}\, p^2 \alpha_{\ell_2} + a_{33}\, p^2 \alpha_{\ell_3} + c_{31}\, \alpha_{\ell_1} + c_{32}\, \alpha_{\ell_2} + c_{33}\, \alpha_{\ell_3} +$$

$$+\, f_3''\, p\, \alpha_{\ell_3} + \left| T_{d.f.}'' \right|\, \text{sign } p\ \alpha_{\ell_3} \pm T_{d_3} = T_{\ell_3}\,,$$

$$(24)$$

where

$$a_{11} = m_1 \tau_1^2 + m_2 (\ell_1 + \tau_2)^2 + m_3 (\ell_1 + \ell_2 + \ell_3)^2\,;$$

$$a_{22} = m_2 \tau_2^2 + m_3 (\ell_2 + \ell_3)^2\,;$$

$$a_{33} = m_3 \ell_3^2\,;$$

$$a_{12} = m_2 \tau_2 (\ell_1 + \tau_2) + m_3 [\,(\ell_2 + \ell_3)^2 + \ell_1 \ell_2 + \ell_1 \ell_3\,]\,;$$

$$a_{13} = m_3 \ell_3 (\ell_1 + \ell_2 + \ell_3)\,;$$

$$a_{23} = m_3 \ell_3 (\ell_2 + \ell_3)\,;$$

$$(25)$$

$$c_{11} = m_1\, g\, \tau_1 + m_2\, g\, (\ell_1 + \tau_2) + m_3\, g\, (\ell_1 + \ell_2 + \ell_3)\,;$$

$$c_{22} = m_2\, g\, \tau_2 + m_3\, g\, (\ell_2 + \ell_3)\,;$$

$$c_{33} = m_3\, g\, \ell_3\,;$$

$$c_{12} = m_2\, g\, \tau_2\,;$$

$$c_{13} = m_3\, g\, \ell_3\,;$$

$$c_{23} = m_3\, g\, \ell_3\,.$$

Taking into consideration the expressions (5) and (6) for T_{ej}, the system of differential equations (24) can be represented by the diagram of Fig. 4 where the notation (25) is used for brevity.

      Using the above procedure, one can investigate not only the plane dynamics of a mechanism, but spatial also, i.e. a slave which has rotation in the shoulder or elbow [1].

Figure 3a represent a kinematic diagram of three-degree of freedom with KDJA slave where the mechanical transmission is performed by cables, and Fig. 3b, a kinematic diagram of a slave where the motion from the actuator is transmitted to an appropriate degree of freedom via tabular shafts and gears.

Analysis of these mechanisms leads to the conclusion that they are similar to the mechanisms of the pantograph type. Fig. 3a represents a simplified kinematic diagram of such an analog with the notation as in fig. 2c.

Assuming that the mass of associated links is concentrated in the points M_1, M_2; M_3 while the torques T_{ej} are reduced to associated joints, the dynamics of such a mechanism can be expressed by Lagrangian equations of the second kind.

The angles α_{ℓ_1} ; α_{ℓ_2} ; α_{ℓ_3} between appropriate elements and axes of the fixed system of coordinates X and Y are assumed to be (Fig. 3c) the generalized coordinates of the system.

Solving eq. (20) and taking into consideration that α_{ℓ_1} = q_1; α_{ℓ_2} = q_2 ; α_{ℓ_3} = q_3 , the dynamics of a slave can be described by the system of differential equations (24) where C_{jk} = 0 while other coordinates are given as

(26)
$$a_{11} = m_1 \tau_1^2 + (m_2 + m_3)\ell_1^2 \ ;$$
$$a_{22} = m_2 \tau_2^2 + m_3 \ell_2^2 \ ;$$
$$a_{33} = m_3 \ell_3^2 \ ;$$
$$a_{12} = m_2 \ell_1 \tau_2 + m_3 \ell_1 \ell_2 \ ;$$
$$a_{13} = m_3 \ell_1 \ell_3 \ ;$$
$$a_{23} = m_3 \ell_2 \ell_3 \ ;$$
$$c_{11} = \tau_1 m_1 g + \ell_1 g(m_1 + m_3) \ ;$$
$$c_{22} = \ell_2 m_2 g + \ell_3 m_3 g \ .$$

The diagram associated with KDJA slaves of robot and manipulator is similar to that of KIJA slaves (Fig. 4) where C_{jk} = 0.

This procedure leads to analytical and structural representation of robot and manipulator slaves as control objects. This representation can be useful in both analytical studies and computer simulation of the dynamics inherent in robot and manipulator control systems.

REFERENCES

[1] КУЛЕШОВ В.С.,ЛАКОТА Н.А.- Динамика систем управления манипуляторами, Изд-во "Энергия", Москва, 1971.

[2] ЧЕСТНАТ Г.,МАЙЕР Р.- Проектирование и расчет следящих систем и систем регулирования.Ч.I и ч.II, Госэнергоиздат, 1959.

[3] ШВЕДОВ В.Н.,КУЛЕШОВ В.С.,ЛАКОТА Н.А.- Структурное аналитическое представления исполнительного органа манипулятора как объекта регулирования, "Известия ВУЗов Машиностроение", Москва, 1970,№ 9.

[4] ЛОЙЦЯНСКИЙ А.Г.,ЛУРЬЕ А.И.- Теоретическая механика, т.II, Москва, 1973.

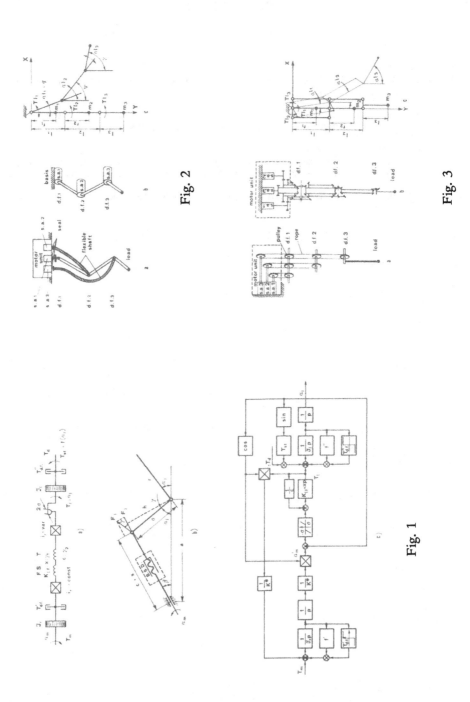

Fig. 2

Fig. 3

Fig. 1

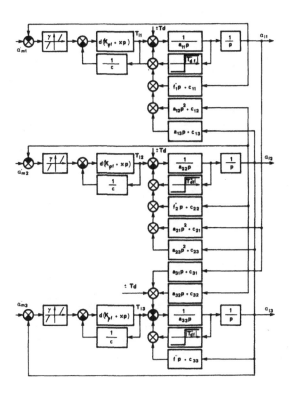

Fig. 4

KINEMATICS AND RATE CONTROL OF THE RANCHO ARM

Maynard L. MOE
University of Denver
Denver, Colorado, U.S.A.

(*)

Summary

The relative position equations for the Rancho arm are developed using the Denavit-Hartenberg representation. Equations for coordinated rate control of the wrist are derived and problems introduced by the mechanical structure of the arm and adoption of rate control are described. These problems include:

1. Nonuniqueness of the rate control equations caused by an extra degree-of-freedom in the mechanical structure.

2. Transformation singularities caused by arm position.

3. Loss of coordinated motion when arm joints reach a mechanical limit.

4. Loss of coordinated motion due to velocity saturation of one or more joints.

Approaches to the solution of these problems are presented.

(*) All figures quoted in the text are at the end of the lecture

1. Introduction

Control of multi-degree-of-freedom teleoperators is of increasing importance in industry, the space program and rehabilitation medicine [8,9,10]. Because of the complexity of the control task there is an opportunity for many creative contributions. A great need exists for improvements in visual displays, tactile sensing, tactile feedback, and control transducers since these affect the communication flow between man and the teleoperator. In many cases the performance of the man-machine system can be improved if a computer is available to process the communication signals. For example, by proper processing it is possible to permit the man to give commands in a coordinate system different than that defined by the degrees-of-freedom of the teleoperator. The development of kinematic representations and rate control strategies which permit an operator to use a convenient coordinate system is the subject to be discussed.

There are two basic approaches to the design of control systems for teleoperators. One is position control where the operator defines the position of the teleoperator and thus specifies the joint angles. The other is rate control where the operator specifies the direction and rate of motion and thus specifies the angular rate for each joint. Rate and position control each have unique properties which may make one or the other most desirable in a particular application. For control of an orthotic device such as the Rancho arm, rate control is used because smooth control can be obtained from control sites with limited dynamic range.

In many systems the type of control used can be obscured by internal processing of signals. For example, input rate signals may be integrated to obtain position signals. Also position signals may be differenced to obtain an error signal which determines a rate of motion for a particular joint [20].

2. Rate Control

The position and orientation of an object in space can be completely specified by six degrees-of-freedom. A complete teleoperator usually has more than six degrees-of-freedom. Complete volitional control of a manipulator requires as many command signals as degrees-of-freedom. Since it is usually inconvenient for the operator to directly command each degree-of-freedom of the teleoperator, some coordinate conversion facility is introduced to permit the commands to be given in a more natural coordinate system [5,14,21]. These coordinate systems can be either fixed in space, moving with the manipulator, such as a hand oriented system, or moving with some external device such as TV camera or the workpiece.

Let the vector, $\underset{\sim}{x}$, denote a set of command variables in some coordinate system or systems. The choice of command variables discussed here will be those for which each command variable has the relation

$$x_i = f_i(\underset{\sim}{\theta}) \tag{1}$$

where, $\underset{\sim}{\theta}$, is the vector representing the degrees-of-freedom of the teleoperator. Differentiating $\underset{\sim}{x}$ with respect to time we obtain

$$\frac{dx}{dt} = \dot{\underset{\sim}{x}} = \underset{\sim}{B}(\underset{\sim}{\theta})\,\dot{\underset{\sim}{\theta}} \tag{2}$$

where $\underset{\sim}{B}(\underset{\sim}{\theta})$ is the Jacobian, i.e., the elements b_{ij}, of $\underset{\sim}{B}$ satisfy the relation

$$b_{ij} = \frac{\partial f_{i-1}}{\partial \theta_{j-1}} \qquad \begin{array}{l} 0 \leqslant i \leqslant n-1 \\ 0 \leqslant j \leqslant n-1 \end{array} \tag{3}$$

where n is the dimension of $\underset{\sim}{x}$ and m the dimension of θ.

One of the advantages of rate control evident in equation (2) is that it is linear at each point in the teleoperator space. Thus, the rate of motion for each degree-of-freedom of the teleoperator can be obtained from

$$\dot{\underset{\sim}{\theta}} = \underset{\sim}{B}^{-1}(\underset{\sim}{\theta})\dot{\underset{\sim}{x}}\,, \tag{4}$$

whenever the inverse exists. Some approaches to the problem when the inverse does not exist will be discussed later.

The first step in the development of rate control is the derivation of the kinematic equations for the system as represented by equation (1) and then the evaluation of the rate equations represented by equation (3). To illustrate the process these steps will be taken using the Rancho arm as the teleoperator.

3. The Kinematics of the Rancho Arm

The Rancho arm is an electrically powered, seven degree-of-freedom splint structure designed to assist in the rehabilitation of severely disabled patients [1]. This arm, sometimes modified, has been used in numerous teleoperator studies [6,13,14,15,16,18,19,24]. The Rancho arm is shown in Figure 1 and its degrees-of-freedom are shown in Figure 2.

The position and orientation of each segment of the linkage structure of the Rancho arm can be descibed using the Denavit-Hartenberg [3,7] representation of coordinate frames in jointed mechanisms. This representation has also been used by Kinzel[11,12] and by Whitney [22,23]. Each of these authors used a slightly different definition of the parameters involved. The notation to be used here has been chosen to follow most closely that of Whitney since he was describing remote manipulators. The structure of the Rancho arm will require an addition to the notation previously used.

Coordinate frames j and j +1 are shown in Figure 3. The Z-axis in each frame is assumed to be the axis of rotation. θ_j is the joint angle, positive in the right hand sense about ^jZ. ^{j+1}X is chosen to be collinear with the common normal between ^jZ and ^{j+1}Z. a_j is the length of the common normal, positive in the direction of ^{j+1}X. a_j is the angle between ^jZ and ^{j+1}Z, positive in the right hand sense about the common normal. S_j is the value of ^jZ at which the common normal intersects ^jZ. When $\theta_j = 0$ then ^jX and ^{j+1}X are parallel and in the same direction. O_j is the origin of the frame j and ^jY is chosen to complete a right hand coordinate system.

If a point in frame j is represented by

$$(5) \qquad \begin{bmatrix} X \\ Y \\ Z \\ 1 \end{bmatrix}_j = {}^j\underline{P}$$

and the same point in frame j + 1 by

$$(6) \qquad \begin{bmatrix} X \\ Y \\ Z \\ 1 \end{bmatrix}_{j+1} = {}^{j+1}\underline{P}$$

then these points are related by

$$(7) \qquad {}^j\underline{P} = {}^j\underline{\underline{A}}_{j+1} \, {}^{j+1}\underline{P}$$

where

$$
{}^{j}_{\sim}A_{j+1} =
\begin{bmatrix}
c\theta_j & -c\alpha_j\, s\theta_j & s\alpha_j\, s\theta_j & a_j\, c\theta_j \\
s\theta_j & c\alpha_j\, c\theta_j & -s\alpha_j\, c\theta_j & a_j\, s\theta_j \\
0 & s\alpha_j & c\alpha_j & S_j \\
0 & 0 & 0 & 1
\end{bmatrix}
\tag{8}
$$

in which $c\alpha_j = \cos\alpha_j$, $s\alpha_j = \sin\alpha_j$, $c\theta_j = \cos\theta_j$ and $s\theta_j = \sin\theta_j$.

In the case when ${}^{j+1}Z$ is parallel to ${}^{j}X$, which occurs when three consecutive perpendicular axes of rotation intersect at a common point, then equation (8) is no longer valid. In this case, ${}^{j}Y$ will be assumed to be in the direction of ${}^{j+1}X$ when $\theta_j = 0$ as shown in Figure 4. For this case ${}^{j}_{\sim}A_{j+1}$ becomes

$$
{}^{j}_{\sim}A_{j+1} =
\begin{bmatrix}
-s\theta_j & -c\alpha_j\, c\theta_j & s\alpha_j\, c\theta_j & a_j\, s\theta_j \\
c\theta_j & -c\alpha_j\, s\theta_j & s\alpha_j\, s\theta_j & a_j\, c\theta_j \\
0 & s\alpha_j & c\alpha_j & S_j \\
0 & 0 & 0 & 1
\end{bmatrix}
\tag{9}
$$

Thus, equation (8) is used when ${}^{j+1}X$ is in the direction of ${}^{j}X$ and equation (9) used when ${}^{j+1}X$ is in the direction of ${}^{j}Y$ (when $\theta_j = 0$). In equations (8) and (9) the upper left 3 x 3 partition expresses the orientation of frame $j + 1$ with respect to frame j, while the upper right 3 x 1 partition gives the location of the origin of frame $j + 1$ with respect to frame j. In each case the relation between successive coordinate systems is seen to depend on four parameters a_j, S_j, α_j and θ_j .

Using this notation to represent the Rancho electric arm, a reference position is chosen with the arm extended straight forward (perpendicular to the frontal plane of the body) with the palm of the hand down and the origin at the intersection of the two axes of rotation of the shoulder joint. The coordinate system is shown in Figure 5. From this reference position a positive rotation of θ_0 moves the wrist to the right; a positive rotation of θ_1 raises the wrist; a positive rotation of θ_2 rotates the arm clockwise when viewed from the shoulder; a positive rotation of θ_3 flexes the elbow raising the hand; a positive rotation of θ_4 rotates the hand clockwise when viewed from the elbow; and a positive rotation of θ_5 raises the fingers. The 6th frame is not associated with a joint of the Rancho arm but can be used to define the position of the fingertips.

The values of a, S, and α for each coordinate frame of the Rancho arm are given in Table I. Using these values

Table I

frame	a	S	α
0	0	0	$-\pi/2$
1	0	0	$+\pi/2$
2	0	L_1	$-\pi/2$
3	0	0	$+\pi/2$
4	0	L_2	$-\pi/2$
5	L_3	0	0

the transformation matrices become

$$(10) \quad {}^{0}\underset{\sim}{A_1} = \begin{bmatrix} c\theta_0 & 0 & -s\theta_0 & 0 \\ s\theta_0 & 0 & c\theta_0 & 0 \\ 0 & -1 & 0 & 0 \\ 0 & 0 & 0 & 1 \end{bmatrix}$$

$$(11) \quad {}^{1}\underset{\sim}{A_2} = \begin{bmatrix} s\theta_1 & 0 & c\theta_1 & 0 \\ c\theta_0 & 0 & s\theta_1 & 0 \\ 0 & 1 & 0 & 0 \\ 0 & 0 & 0 & 1 \end{bmatrix}$$

$$(12) \quad {}^{2}\underset{\sim}{A_3} = \begin{bmatrix} c\theta_2 & 0 & s\theta_2 & 0 \\ s\theta_2 & 0 & c\theta_2 & 0 \\ 0 & -1 & 0 & L_1 \\ 0 & 0 & 0 & 1 \end{bmatrix}$$

$$(13) \quad {}^{3}\underset{\sim}{A_4} = \begin{bmatrix} c\theta_3 & 0 & s\theta_3 & 0 \\ s\theta_3 & 0 & -c\theta_3 & 0 \\ 0 & 1 & 0 & 0 \\ 0 & 0 & 0 & 1 \end{bmatrix}$$

$$(14) \quad {}^{4}\underset{\sim}{A_5} = \begin{bmatrix} c\theta_4 & 0 & -s\theta_4 & 0 \\ s\theta_4 & 0 & c\theta_4 & 0 \\ 0 & -1 & 0 & L_2 \\ 0 & 0 & 0 & 0 \end{bmatrix}$$

$$(15) \quad {}^{5}\underset{\sim}{A_6} = \begin{bmatrix} c\theta_5 & -s\theta_5 & 0 & L_3 c\theta_5 \\ s\theta_5 & c\theta_5 & 0 & L_3 s\theta_5 \\ 0 & 0 & 1 & 0 \\ 0 & 0 & 0 & 1 \end{bmatrix}$$

These matrices can be used to relate any coordinate frame to another. For example, the position and orientation of the wrist (frame 5) relative to the shoulder (frame 0) is given by

$$^0\underset{\sim}{A_5} = {}^0\underset{\sim}{A_1}\ {}^1\underset{\sim}{A_2}\ {}^2\underset{\sim}{A_2}\ {}^3\underset{\sim}{A_4}\ {}^4\underset{\sim}{A_5} \tag{16}$$

By further abbreviating $c\theta_j$ to c_j and $s\theta_j$ to $s_j (i = 0,1,2,3,4)$, the elements, a_{ij}, of $^0A_5$ can be written

$$
\begin{aligned}
a_{11} &= -c_0\, s_1\, c_2\, c_3\, c_4 - s_0\, s_2\, c_3\, c_4 - c_0\, c_1\, s_3\, c_4 + c_0\, s_1\, s_2\, s_4 - s_0\, c_2\, s_4 \\
a_{12} &= c_0\, s_1\, c_2\, s_3 + s_0\, s_2\, s_3 - c_0\, c_1\, c_3 \\
a_{13} &= c_0\, s_1\, c_2\, c_3\, s_4 + s_0\, s_2\, c_3\, s_4 + c_0\, c_1\, s_3\, s_4 + c_0\, s_1\, s_2\, c_4 - s_0\, c_2\, c_4 \\
a_{14} &= L_2\,(c_0\, c_1\, c_3 - c_0\, s_1\, c_2\, s_3 - s_0\, s_2\, s_3) + L_1\, c_0\, c_1 \\
a_{21} &= -s_0\, s_1\, c_2\, s_3\, c_4 + c_0\, s_2\, s_3\, c_4 - s_0\, c_1\, s_3\, c_4 + s_0\, s_1\, s_2\, s_4 + c_0\, c_2\, s_4 \\
a_{22} &= s_0\, s_1\, c_2\, s_3 - c_0\, s_2\, s_3 - s_0\, c_1\, c_3 \\
a_{23} &= s_0\, s_1\, c_2\, c_3\, s_4 - c_0\, s_1\, c_3\, s_4 + s_0\, c_1\, s_3\, s_4 + s_0\, s_1\, s_2\, c_4 + c_0\, c_2\, c_4 \\
a_{24} &= L_2\,(s_0\, c_1\, c_3 - s_0\, s_1\, c_2\, s_3 + c_0\, s_2\, s_3) + L_1\, s_0\, c_1 \\
a_{31} &= -c_1\, c_2\, c_3\, c_4 + s_1\, s_3\, c_4 + c_1\, s_2\, s_4 \\
a_{32} &= c_1\, c_2\, s_3 + s_1\, c_3 \\
a_{33} &= c_1\, c_2\, c_3\, s_4 - s_1\, s_3\, s_4 + c_1\, s_2\, c_4 \\
a_{34} &= -L_2\,(c_1\, c_2\, s_3 + s_1\, c_3) - L_1\, s_1 \\
a_{41} &= a_{42} = a_{43} = 0 \\
a_{44} &= 1
\end{aligned}
\tag{17}
$$

These relations can be used to determine some possible control equations for the Rancho arm. The upper left 3 x 3 partition gives the orientation of the wrist (frame 5) with respect to frame 0 and the upper right 3 x 1 partition gives the location of the wrist (the origin of frame 5).

4. Rate Control of the Rancho Arm

Assume rate control of wrist position is desired. Let X_w, Y_w, and Z_w be the Cartesian coordinates of the wrist. If we let these coordinates in frame 0 be the command variables we have

(18)
$$
\begin{bmatrix} x_0 \\ x_1 \\ x_2 \end{bmatrix} = \begin{bmatrix} X_w \\ Y_w \\ Z_w \end{bmatrix} = \begin{bmatrix} f_0(\underline{\theta}) \\ f_1(\underline{\theta}) \\ f_2(\underline{\theta}) \end{bmatrix} = \begin{bmatrix} a_{14} \\ a_{24} \\ a_{34} \end{bmatrix}
$$

where $(\underline{\theta})$ can be reduced to

(19)
$$
\underline{\theta} = \begin{bmatrix} \theta_0 \\ \theta_1 \\ \theta_2 \\ \theta_3 \end{bmatrix}
$$

since a_{14}, a_{24}, and a_{34} are only functions of these degrees-of-freedom. Differentiating equation (18) we obtain

(20)
$$
\begin{bmatrix} \dot{x}_0 \\ \dot{x}_1 \\ \dot{x}_2 \end{bmatrix} = \begin{bmatrix} b_{11} & b_{12} & b_{13} & b_{14} \\ b_{21} & b_{22} & b_{23} & b_{24} \\ b_{31} & b_{32} & b_{33} & b_{34} \end{bmatrix} \begin{bmatrix} \dot{\theta}_0 \\ \dot{\theta}_1 \\ \dot{\theta}_2 \\ \dot{\theta}_3 \end{bmatrix}
$$

where each b_{ij} can be evaluated using equation (3) to give

(21)
$$
\begin{aligned}
b_{11} &= L_2 (s_0 s_1 c_2 s_3 - s_0 c_1 c_3 - c_0 s_2 s_3) - L_1 s_0 c_1 \\
b_{12} &= -L_2 (c_0 s_1 c_3 + c_0 c_1 c_2 s_3) - L_1 c_0 s_1 \\
b_{13} &= L_2 (c_0 s_1 s_2 s_3 - s_0 c_2 s_3) \\
b_{14} &= -L_2 (c_0 c_1 s_3 + c_0 s_1 c_2 c_3 + s_0 s_2 c_3) \\
b_{21} &= L_2 (c_0 c_1 c_3 - c_0 s_1 c_2 s_3 - s_0 s_2 s_3) + L_1 c_0 c_1 \\
b_{22} &= -L_2 (s_0 s_1 c_3 + s_0 c_1 c_2 s_3) - L_1 s_0 s_1 \\
b_{23} &= L_2 (s_0 s_1 s_2 s_3 + c_0 c_2 s_3) \\
b_{24} &= L_2 (c_0 s_2 c_3 - s_0 c_1 s_3 - s_0 s_1 c_2 c_3) \\
b_{31} &= 0 \\
b_{32} &= L_2 (s_1 c_2 s_3 - c_1 c_3) - L_1 c_1 \\
b_{33} &= L_2 c_1 s_2 s_3 \\
b_{34} &= L_2 (s_1 s_3 - c_1 c_2 c_3)
\end{aligned}
$$

where it might be noted that $b_{11} = -a_{24}$ and $b_{21} = a_{14}$ which could be used to simplify the calculations in some situations where position is also being calculated.

Equations (20) and (21) give the relationships needed for rate control of wrist position in a Cartesian coordinate system. Similar equations can be developed for spherical coordinates, hand oriented coordinates, as well as the attitude of the hand and/or arm. Of course, these equations become more complex as coordinate systems involving additional degrees-of-freedom are used. Equations (20) and (21) are typical of those obtained and can be used to illustrated some problems related to coordinated rate control.

5. Problems Encountered

Four basic problems encountered in the use of equation (20) to solve for the joint angle rates are:

1. There are 4 joint rates and only three rate commands so an extra degree-of-freedom is present which results in a nonsquare $\underset{\sim}{B}$ matrix.

2. In some positions the $\underset{\sim}{B}$ matrix may not have full rank.

3. One of the joints may be at a limit and cannot move in the desired direction.

4. One ormore joints may not be able to go as fast as the solution to equation (20) requires.

Each of these problems and possible solutions will be described below.

Since the B matrix for equation(20) is not square, equation (4) cannot directly be used to solve for $\dot{\theta}$. One approach which can be taken is the use of generalized inverses [2,17]. When $\underset{\sim}{B}$ has more columns than rows the solution of equation (2) which minimizes $\underline{\dot{\theta}}^T \underline{\dot{\theta}}$ is given by

$$\underline{\dot{\theta}} = \underset{\sim}{B}^T (\underset{\sim}{B}\underset{\sim}{B}^T)^{-1} \underline{\dot{x}} . \qquad (22)$$

This can also be generalized to minimize $\dot{\theta}^T Q \dot{\theta}$ by setting

$$\underline{\dot{\theta}} = \underset{\sim}{Q}^{-1} \underset{\sim}{B}^T (\underset{\sim}{B}\underset{\sim}{Q}^{-1}\underset{\sim}{B}^T)^{-1} \underline{\dot{x}} \qquad (23)$$

where $\underset{\sim}{Q}$ is a symmetric positive definite weighting matrix. Obviously, equations (22) and (23) require more computation than would be needed if $\underset{\sim}{B}$ were square and nonsingular permitting equation (4) to be used.

A second approach to the nonsquare matrix problem is to make it square. If a single degree-of-freedom is held constant and the remaining 3 degrees-of-freedom used to solve for the joint rates, then one column of $\underset{\sim}{B}$ can be eliminated and equation (4) used. For the Rancho arm it is most convenient to assume θ_2 (humeral rotation) is held fixed; which effectively determines the elevation of the elbow relative to a line from the shoulder to the wrist. Of course, θ_2 can be changed at any time to vary the elbow elevation if the wrist position is held fixed. This approach has been used very effectively in previous research [15,16]. A third approach using statistical procedures to determine elbow position and thus eliminate the extra degree-of-freedom has been reported [13].

To illustrate the second problem let us analyze the singularities of the generalized inverse approach as well as the approach which simply strikes out a column of $\underset{\sim}{B}$. One might expect the generalized inverse approach to have fewer singularities than the approach which strikes out the third column of $\underset{\sim}{B}$; but we will see that this is not so. Let $\underset{\sim}{D}$ be the matrix obtained from $\underset{\sim}{B}$ by striking out the third column. It is evident that the rank of $\underset{\sim}{B}$ is at least as large as that of $\underset{\sim}{D}$; we will show they always have equal rank. It can be shown, by setting the determinant to zero, that $\underset{\sim}{D}$ is singular if, and only if, $\theta_3 = 0$. (There is also a singularity at $\theta_3 = \pi$ but the arm cannot reach this position so it is ignored. The same arguments could be used if it were included.) Thus $\underset{\sim}{B}$ and $\underset{\sim}{D}$ have the same rank when $\theta_3 \neq 0$. If we let $\theta_3 = 0$ and $L_a = L_1 + L_2$ in equation (20), then

$$
(24) \quad \underset{\sim}{B}\Big|_{\theta_3 = 0} = \begin{bmatrix} - L_a s_0 c_1 & - L_a c_0 s_1 & 0 & - L_2 (c_0 s_1 c_2 + s_0 s_2) \\ - L_a c_0 c_1 & - L_a s_0 s_1 & 0 & L_2 (c_0 s_2 - s_0 s_1 c_2) \\ 0 & - L_a c_1 & 0 & - L_2 c_1 c_2 \end{bmatrix}
$$

Since the third column disappears $\underset{\sim}{B}$ and $\underset{\sim}{D}$ will have the same rank when $\theta_3 = 0$. Thus $\underset{\sim}{B}$ and $\underset{\sim}{D}$ always have the same rank. Then, since the rank of $\underset{\sim}{B}\underset{\sim}{B}^T$ and $\underset{\sim}{B}\underset{\sim}{Q}^{-1}\underset{\sim}{B}^T$ is the same as the rank of $\underset{\sim}{B}$ it is evident that the generalized inverses used in equations (22) and (23) will have the same singularity problems as encountered when $\underset{\sim}{D}$ is used in place of $\underset{\sim}{B}$ in equation (4). This result is to be expected, since, when θ_3 is very close to zero, θ_2 has very little effect on the position of the wrist so there is little to be gained by including it in the calculations. Of course, θ_2 has a significant impact on the wrist position when θ_3 is close to $\pi/2$ so the generalized inverse procedures may result in lower speeds for some joints than the use of $\underset{\sim}{D}$ and thus may avoid speed saturation of the motors and preserve better coordinated motion. The existence of a singularity is determined by the mechanical structure and cannot be eliminated by the use of a different coordinate system or the generalized inverse. However, it is possible to introduce additional singularities by a poor approach to the solution of the rate equations. For example, if we formed a matrix $\underset{\sim}{G}$ by deleting the second column of $\underset{\sim}{B}$, then $\underset{\sim}{G}$ will be singular when either $\theta_3 = 0$, as shown previously, or when $\theta_1 = -\pi/2$ since

$$
\underset{\sim}{G}\Big|_{\theta_1 = -\frac{\pi}{2}} = L_2 \begin{bmatrix} (-s_0 c_2 s_3 - c_0 s_2 s_3)(-s_0 c_2 s_3 - c_0 s_2 s_3) & (c_0 c_2 c_3 - s_0 s_2 c_3) \\ (c_0 c_2 s_3 - s_0 s_2 s_3)(c_0 c_2 s_3 - s_0 s_2 s_3) & (c_0 s_2 c_3 + s_0 c_2 c_3) \\ 0 & 0 & -s_3 \end{bmatrix}
$$

(25)

where the first and second columns are identical. The $\underset{\sim}{G}$ matrix is obviously not as good as $\underset{\sim}{D}$ for control. If the structure can be changed the singularities can be avoided. For example, the addition of telescoping segment in the distal or proximal portion of the Rancho arm would eliminate the singularity in the resulting $\underset{\sim}{B}$ matrix.

In practice it is probably neither necessary nor desirable to complicate the design of a teleoperator by adding one or more degrees-of-freedom to eliminate singularities. One obvious solution is to limit joint motion so the teleoperator cannot be driven to a point where a singularity exists. This solution, while simple, is often too restrictive. A second related solution is to modify the joint angle data so a singular solution does not result. This approach has been used in control of the Rancho arm where angle θ_3 was set to a fixed angle θ_{min} whenever θ_3 was less than θ_{min} [15,16]. Since the actual joint angle was a few degrees from the angle used in the equations a degradation in coordinated motion near the singularity resulted. However, the change in performance was not noticeable to the user.

A third solution near the singularity is to adopt a simple strategy such as temporarily using direct joint angle control. Finally, since the singularity results in a $\underset{\sim}{B}$ matrix without full rank it would be possible to eliminate rows of $\underset{\sim}{B}$ until a matrix of full rank is obtained. Since each row of $\underset{\sim}{B}$ defines the effect of $\underset{\sim}{\dot\theta}$ on motion along one axis the elimination of a row turns control along this axis into a "don't care" condition. That is, there is no control of motion in the dimension controlled by the corresponding input. This results in a loss of coordination but further research is needed to determine how serious it is.

It is well to note that all solutions result in a departure from coordinated motion near the singularity. The singularity is simply an indication that the desired coordinated motion is impossible but not that motion is impossible. The particular solution adopted will depend on the intended application of the teleoperator.

A third problem is encountered when one or more joints of the teleoperator reach a limit. If further motion toward the limit is required the goal of coordinated motion is subverted. This problem is quite severe in orthotic applications because of the limited range of motion of many disabled persons. One possible solution is to ignore the problem by stopping each joint at its limit but permitting the other joints to continue moving. The result can be quite unnerving to the user as it often appears to him that he has lost control. A second approach is to stop all joints when any joint is moving against a limit. This preserves coordinated motion but often severely limits the range of motion of the teleoperator. Also, since the inputs are not joint commands, it is often difficult for the operator to determine a command that will result in motion away from the limit. The lack of response to an input command that is illegal because it requires motion into a limit is very annoying and tends to lead the operator to conclude that some system failure occurred, such as loss of power.

A third solution to the problem is to eliminate from the solution those joints which are being driven into limits. This can be done by deleting the corresponding columns of $\underset{\sim}{B}$. The previously discussed methods can then be used to solve for the remaining joint angle rates as long as the number of columns is greater than or equal to the number of rows. If the number of columns is less than the number of rows then coordinated motion in the desired direction is impossible. Motion in a direction which minimizes the sum of squares of residuals, $J = \underline{r}^T \underline{r}$, where

$$(26) \qquad\qquad \underline{r} = \underline{\dot{x}} - \underset{\sim}{B}_r \underline{\dot{\theta}}_r$$

in which $\underset{\sim}{B}_r$ is the reduced $\underset{\sim}{B}$ matrix and $\underline{\dot{\theta}}_r$ is obtained from $\underline{\dot{\theta}}$ by deleting the joints being driven into a limit, is possible and can be obtained by using 17

$$(27) \qquad\qquad \underline{\dot{\theta}}_r = (\underset{\sim}{B}_r^T \underset{\sim}{B}_r)^{-1} \underset{\sim}{B}_r^T \underline{\dot{x}}.$$

This will give motion as close, in a mean-square sense, to the desired direction as possible.

However, when a joint is at a limit it is still necessary to permit motion away from the limit. Thus, before a joint is eliminated from the calculations it is necessary to solve for the rates using the full $\underset{\sim}{B}$ matrix to determine joint directions and if motion away from a limit is indicated that joint will remain in the computation even when at a limit. Therefore, when a joint is at a limit, the rate equations may have to be solved twice which essentially doubles the computational effort required. Of course the elements of the $\underset{\sim}{B}$ matrix do not have to be recalculated.

A fourth problem exists when one or more joints may not be able to go as fast as necessary to maintain coordinated motion. Use of the generalized inverse may help avoid speed saturation, however, since operators will always tend to use a system to its maximum capabilities, speed saturation will occur. Fortunately, for this problem there is a fairly simple and effective solution. Given that the maximum speed for each joint is known, it is fairly simple to examine the computed rates to see if any exceed its specified maximum. If one or more do, the output rates can be scaled down until they are all within their maximum. Since the rate equations are linear the resulting motion will still be coordinated. The operator is not usually aware that such scaling is taking place since he can see that at least one of the joints is moving quite rapidly. If the scaling is not done the loss of coordination can be quite noticeable.

All of the problems discussed are interrelated to some extent and the procedures used to solve one problem will affect the solution to others. Unfortunately, the use of generalized inverses, which is theoretically very appealing, requires much more computation that some ad hoc solutions, such as eliminating columns or rows of $\underset{\sim}{B}$ to make it square. Since these computations introduce delays into the system, and it is well known that performance of a human operator tends to decrease as delay is increased, it is important to minimize the computation time [4]. In some applications, where the speed of the teleoperator is not great, it is possible to do most of the computations involved in solving the rate equations in a background processing mode and to multiply the input $\dot{x}$ by the appropriate matrix several times before a new $\underset{\sim}{B}$ is computed and used. Since the arm is moving slowly the fact that old position data is being used for some of the rate calculations will not seriously reduce the coordination of motion and will result in a very quick response to input commands. For example, input commands could be sampled and new rates computed every 50 milliseconds while the generalized inverse is recalculated every 200 milliseconds.

Finally, it must be emphasized that in any teleoperator application the investment in computational capability needs to be balanced against the investment in the teleoperator structure, motor drive circuits, position sensors, tactile sensors and the viewing system. Use of the Rancho arm for rehabilitation indicates a low cost system is needed. One configuration of the system using ocular transducers to obtain rate commands is shown in Figure 6. Some details of this system, which used open-loop speed control of the motors, have been published elsewhere [16].

6. Conclusions

Kinematic relations were derived for the Rancho arm and used to obtain a set of equations which can be used to provide coordinated rate control. Several problems in coordinated rate control were described and several solutions presented. Since coordinated control is introduced to simplify the control task, care must be taken to insure that problems, such as time delay, introduced by this control do not prevent effective implementation. Although coordinated control cannot be used to overcome limited capabilities in range of motion, speed of operation, etc., it can be used to more effectively utilize the capability of the teleoperator by improving man-machine communication.

7. Acknowledgements

This research was supported in part by Grant AM-10763 from the National Institutes of Health, Department of Health Education and Welfare, Washington, D.C. The assistance of John Schwartz and many students in the implementation of the system showhn in Figure 6 is appreciated.

REFERENCES

[1] Allen, J.R., et al, "The Rancho Electric Arm," Record of the Third Annual Rocky Mountain Bioengineering Symposium, pp. 79-82, May 1966.

[2] Boullion, T.L., and Odell, P.L.,"Generalized Inverse Matrices," Wiley-Interscience, New York, 1971.

[3] Denavit, J., and Hartenberg, R.S., "A kinematic Notation for Lower-Pair Mechanisms Based on Matrices," ASME J. Applied Mechanics, Vol. 23, pp. 215-221, June 1955.

[4] Ferrell, W.R., "Remote Manipulation with Transmission Delay," IEEE Trans. on Human Factors in Electronics, pp. 24-32, September 1965.

[5] Gavrilović, M.M., and Marić, M.R., "An Approach to the Organization of the Artificial Arm Control," Proc. of the Third International Symposium on External Control of Human Extremities, pp. 307-322, Dubrovnik, Yugoslavia, August 1969.

[6] Greeb, F.J., "Equations of motion for Control of an Upper Extremity Splint Structure," M.S. Thesis, Department of Electrical Engineering, University of Denver, May 1970.

[7] Hartenberg, R.S., and Denavit, J., "Kinematic Synthesis of Linkages," McGraw-Hill, New York, 1964.

[8] Johnsen, E.G., and Corliss, W.R., "Teleoperators and Human Augmentation," NASA SP-5047, 265 pages, December 1967.

[9] Johnsen, E.G., and Magee, C.B., (editors), "Advancements in Teleoperator Systems," Report of a Colloquium held at the University of Denver, Denver, Colorado, February 1969, NASA SP-5081, 1970.

[10] Johnsen, E. G., and Corliss, W.R.,"Human Factors Applications in Teleoperator Design and Operation", Wiley-Interscience, 1971.

[11] Kinzel, G.L., Hall, A.S., and Hillberry, B.M., "Measurements of the Total Motion Between Two Body Segments-I. Analytical Development," J. Biomechanics, Vol. 5, pp. 93-109, 1972.

[12] Kinzel, G.L., et al., "Measurement of the Total Motion Between Two Body Segments-II. Description of Application," J. Biomechanics, Vol. 5, pp. 283-293, 1972.

[13] Lawrence, P.D., and Lin, W.C., "Statistical Decision Making in the Real-Time Control of an Arm Aid for the Disabled," IEEE Transactions on Systems, Man, and Cybernetics, Vol. SMC-2, No. 1, pp. 35-42, January 1972.

[14] Moe, M.L., and Schwartz, J.T., "A Coordinated, Proportional Motion Controller for an Upper-Extremity Orthotic Device," Proc. of the Third International Symposium on External Control of Human Extremities, pp. 295-305, Dubrovnik, Yugoslavia, August 1969.

[15] Moe, M.L., and Schwartz, J.T., "Ocular Control of the Rancho Electric Arm," Proc. of the Fourth International Symposium on External Control of Human Extremities, Dubrovnik, Yugoslavia, August 1972.

[16] Moe, M.L., and Schwartz, J.T., "Control of the Rancho Electric Arm," Proc. 1972 Fall Joint Computer Conference, pp. 1081-1087, December 1972.

[17] Nobel, B.,"Applied Linear Algebra", Prentice Hall, Inc., Englewood Cliffs, N.J., 1969.

[18] Schwartz, J.T., Moe, M.L. and Hedberg, C.A., "A Coordinated Motion Controller for an Electric Arm Orhtesis-Preliminary Report," Fifth Annual Rocky Mountain Bioengineering Symposium Record, pp. 44-48, May 1968.

[19] Schwartz, J.T., and Moe, M.L., "A Coordinated Motion Controller for the Rancho Electric Arm," Record of the Sixth Annual Rocky Mountain Bioengineering Symposium, pp. 85-86, May 1969.

[20] Vukobratović, M., and Juricić, D., "Note on a Way of Moving the Artificial Upper Extremity," IEEE Trans. on Biomechanical Engineering, Vol. BME-16, No. 2, pp. 113-115, April 1969.

[21] Whitney, D.E., "Resolved Motion Rate Control of Manipulators and Human Prostheses," IEEE Transactions on Man-Machine Systems, Vol. MMs-10, No. 2, pp. 47-54, June 1969.

[22] Whitney, D.E., "The Mathematics of Coordinated Control of Prosthetic Arms and Manipulators." Journal of Dynamic Systems, Measurement and Control, Trans. ASME, Vol. 94, Series G, pp. 303-309, December 1972.

[23] Whitney, D.E., "The Mathematics of Coordinated Control of Prostheses and Manipulators," Proc. of the Fourth International Symposium on External Control of Human Extremities, Dubrovnik, Yugoslavia, August 1972.

[24] Zebo, T.J., et al., "Myoelectric Control of the Rancho Electric Arm," Proc. of the 21st Annual Conference on Engineering in Medicine and Biology, November 1968.

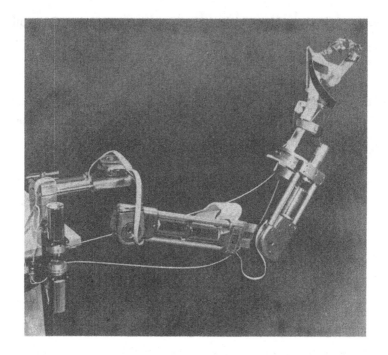

Fig. 1   Rancho electric arm.

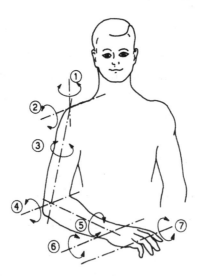

Fig. 2   Degrees-of-freedom of the Rancho arm.

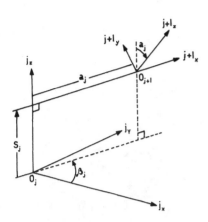

Fig. 3   Definition of parameters
in coordinate system.

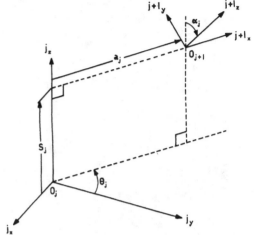

Fig. 4   Coordinate system for case where
^{j+1}x is in the direction of ^iY when
$\theta_j = 0$.

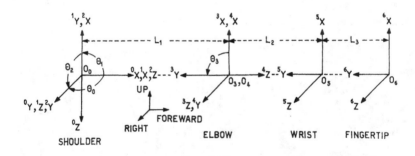

Fig. 5   Coordinate system of the Rancho arm

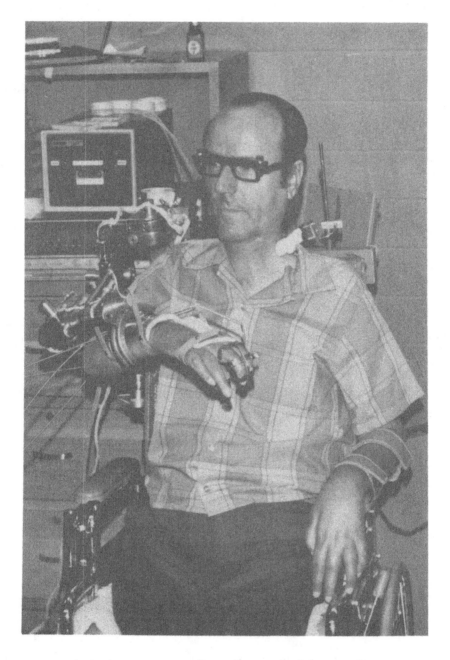

Fig. 6   Patient using coordinated control of the Rancho arm.

THE FORCE VECTOR ASSEMBLER CONCEPT

J.L. NEVINS
C.S. Draper Laboratory
Cambridge, Massachusetts USA

D.E. WHITNEY
Department of Mechanical Engineering
Massachusetts Institute of Technology
Cambridge, Massachusetts USA

(*)

Summary

A factory for making discrete parts and assembling them into products has four principle elements: 1) material processing systems, 2) internal transportation systems, 3) assembly systems, 4) inspection systems. People play a major role in each of these systems, either supervising or actually performing the necessary functions.

To automate such a factory would require development of new systems for assembly and inspection. Considerable automation exists now in the other areas. High volume, fixed configuration assembly systems are available but these solve simple assembly problems only and are not applicable to small production lots or to mixtures of models.

This paper outlines concepts and describes research results for an assembler system organized around force, tactile and displacement information generated at the interface between two pieces which are being assembled. A mechanical arm (or arms), equipped with sensors in the wrist, communicates with a control computer to accomplish tasks such as putting pegs in tightly fitting holes, following edges, and other tasks in which control action is based on information fed back from the task itself.

Alternate methods of organizing such systems are discussed, along with some of the important trade-offs.

(*) All figures quoted in the text are at the end of the lecture.

1. Introduction

In most industrialized nations, the large volume manufacture of discrete parts has been highly automated. They are assembled into substantially identical products by people. High labor costs and the boredom of such jobs exerting pressure to automate the assembly process. It might seem natural to build special purpose machines for assembly similar to those currently in use for material processing and, indeed, a few such machines have been built. Several facts argue against taking this route, however:

1) When people assemble things they also perform vital inspection functions, seeing that the parts are correct and that previous assembly steps have been carried out correctly.

2) There is a growing trend away from identical products and toward products which are specialized for individual customers. While some basic parts will be common to all the versions of a product, many others will be different. This is called the model mix problem.

3) The model mix problem has two main consequences. First, production volumes for each version of the product are relatively small and the stream of incoming orders for product is random. People are adaptable to a random job stream, but only to a limited extent. Thus the second consequence is that assembly mistakes often occur when the wrong part is installed in a subassembly.

This all means that while fixed-configuration assembly machines have limited applicability in such assembly problems, so also people have limited applicability. Current automatic assembly systems solve simple assembly problems and do so by using an open-loop positional control approach. This requires very expensive jigs and fixtures which can be paid for only with high production volume.

As an alternative to fixed automation for assembly, we are investigating flexible (that is, reprogrammable) automated assembly systems. Reprogrammability allows such a system to be taught to assemble different products so that it can be used even after the first product it assembled goes out of production. Also, the system can switch programs in real time and thus can handle a random stream of products it knows how to assemble. Such a system is based on closed loop behavior. That is, it does not require precise knowledge of parts location or that different parts in a series be manufactured identically. Instead, it relies to some extent on information it receives from sensors during assembly itself to guide the parts together and to detect gross errors in parts manufacture or avoid jamming if the parts do not go together correctly due to manufacturing errors. Thus some rudimentary inspection is also performed.

An automated sub-factory containing such assembler systems is indicated schematically in Figure 1. Its basic parts are material processing, intra-factory transfer or transportation, assembly and inspection. Any near-future realization of such a system would employ people at least as supervisors, inspectors, trouble shooters and teachers.

2. Basic System

A basic assembler system as illustrated by Figure 2 consists of a hand (tongs?) and a wrist (the total may be called an end point) plus a mobility system with at least six-degrees of freedom to enable positioning of the end point at any spot and at any orientation in a work volume. Since this may not be practically possible, three systems are shown. Also, for practical assembly problems one or more systems will probably be needed anyway.

In the wrist portion of the end point is an array of force transducers capable of accurately measuring both a rectilinear force vector or its three components plus the three torque components. In addition, tactile and obstacle avoidance sensors are needed as well as a means for determining the displacement or location of the mobility system in the work volume. In addition to the sensor system, the assembler contains computers or special processors capable of making high-speed comparisons between list files of desired force vectors, state point locations, tactile forces and the measured force vectors and displacement data. The processor must also resolve these error signals, based on a predetermined strategy or control algorithm (part of the original list file) to reposition the end point to obtain the desired force vector or displacement.

3. System Organization

How would such an assembler system be organized ? What information would it need to be given prior to assembly (the teaching process). What information would it need during assembly, and how would it use this information ?

One type of system organization is based on very detailed mathematical descriptions for each part and each step in the assembly process. Each piece is defined by a state point vector and rotation matrix telling where it is located and oriented with respect to a reference coordinate system. Descriptions of assembly steps then consist of statements describing the desired relative positions and orientations of a part and some previously assembled parts. Figure 3 contains this kind of description for a common stapler.

Some difficulties with this approach are that much effort must be expended to generate the descriptions, which are unique to each part and are vulnerable to manufacturing variations from one part to another. It is difficult to generalize such descriptions, although some progress has been made (Ref. 1). In addition, there is no direct provision for feedback information during assembly.

Another technique is basically discrete, logical and binary, and is based on MANTRAN (ref. 2). This approach has also been used by Hill and Bliss (ref. 3). It makes use primarily of on-off touch feedback and integrates it with a control program containing statements like "MOVE AHEAD UNTIL ANY TOUCH SENSOR TURNS ON." Discrete move searches and loops which terminate on sensor return can be programmed this way. In situations where the geometry of the environment is known or where tasks can be described as a branching-covering tree structure, such methods are useful. They can be cumbersome in multi-axis situations where a multiplicity of discrete inputs in a particular sequence is being sought, or where the task is not a tree, or where the real time digital processing is too time consuming.

Many tasks we do with our hands are not binary in nature except at points where a failure is being detected. Rather, we are processing more or less continuous variable data more or less continuously in time (relative to the arm's bandwidth). Some types of tasks require that we organize the forces exerted by each of our muscles in order that large force can be exerted at the hand in directions which the environment can easily alter. Opening a door, inserting a peg in a hole or turning a crank are examples. Describing these tasks as requiring one or more joints of the arm to go slack is too narrow a view. Rather, one or more directions of hand motion must go slack or equivalently must yield to forces coming in from the environment. Due to uncertainty concerning the environment's exact relative position and orientation, one cannot predict which joints would need to be slackened, especially in a non-planar arm. It is more relevant to refer such matters to the hand itself, and also easier to visualize.

Thus, the third approach to problems of organizing hand motion and sensor feedback is to monitor a vector of forces and torques at the hand, as well as joint positions and rates and any touch and grip force sensors seemed necessary. Within this approach there seem to be two cases of interest, defining the casuality of command and feedback:

1) command a force and look at the displacements of feedback velocities

2) command a position or a velocity and look at feedback forces

Case 1 is usually appropriate in situations where the hand is moving unopposed by the environment. This requires highly stable control algorithms capable of adapting to changing arm configuration (continuous) and load (intermittant but logically linked to the task). Situations where the arm must behave as a multivariate spring or damper can also be approached this way. Examples include catching moving loads, damping vibrations, and so on. A block diagram of this situation is shown in Figure 4. This diagram is for the case where load inertia dominates. $\underline{F}_d$ is the desired force vector, $\underline{\tau}$ is a vector of joint torques, J^T is the transpose of the Jacobian matrix of the arm (*), and the feedback matrix, if passive, represents damping but could represent other types of impedances instead. $\dot{\underline{x}}$ is a vector of velocities at the hand in hand oriented coordinates. (See Figure 7) $\underline{F}_d$ is also expressed in these coordinates. Communication with and programming of a logical structure such as this would be straightforward and its realization could be in hardware, freeing the control computer for other tasks.

The idea here is to give the arm certain force-velocity characteristics rather than to make it a force-source. The latter is also possible and useful, however. Its structure is shown in Figure 5.

This hookup is a straighforward servo whose logic is actually being generated upstream of $\underline{F}_d$ as a desired force-time trajectory. The feedback matrix controller has no special meaning in this mode except to guarantee that $\underline{F} = \underline{F}_d$. The case where arm inertia is relevant may be treated by using

$$\underline{\tau} = J^T \underline{F} + I_a J^{-1} I_L^{-1} \underline{F}$$

where I_a and I_L are the inertia matrices of the arm (computed in arm coordinates) and the load (computed in load coordinates) respectively. Coriolis forces are ignored in deriving this relationship.

Case 2 is interesting in the situation where the hand and/or its load contact the environment. Here the gross logic of the task is expressed by a desired hand velocity ($\dot{\underline{x}}_d$) but the returning forces ($\underline{F}$) give clues which cause modifications to the desired velocity. A block diagram of the logic flow is shown in Figure 6. Here the choice of the matrix which converts feedback force vector $\underline{F}$ into velocity

---

(*) The Jacobian matrix relates small changes in hand position to small changes in joint angles. See ref. 4 for details.

modifications determines the behavior of the arm. Consider some examples:

> NOTE: For all examples the coordinate frame to be used is shown in Figure 7. This coordinate frame when activated by an undefined mobility system in a rate mode to achieve the desired velocities along the indicated axes, uses a computational scheme which has been called Resolved Motion Rate Control. (ref. 4).

For these examples we will also need to assume that

$$\underline{\dot{x}} = \frac{d}{dt} \begin{vmatrix} \text{Reach} \\ \text{Sweep} \\ \text{Lift} \\ \text{Twist} \\ \text{Tilt} \\ \text{Turn} \end{vmatrix}$$

Example 1: Put a peg in a hole, starting from a point where the peg is in the mouth of the hole but is misaligned.

Set

$$\underline{\dot{x}}_d = \begin{vmatrix} \text{Reach} \\ 0 \\ 0 \\ 0 \\ 0 \\ 0 \end{vmatrix}$$

The Admittance Matrix

$$\begin{bmatrix} 1 & 0 & 0 & 0 & 0 & 0 \\ 0 & -100 & 0 & 0 & 0 & 0 \\ 0 & 0 & -100 & 0 & 0 & 0 \\ 0 & 0 & 0 & -100 & 0 & 0 \\ 0 & 0 & 0 & 0 & -100 & 0 \\ 0 & 0 & 0 & 0 & 0 & -100 \end{bmatrix}$$

This will cause   motions other than reach to occur in accomodation to cross forces and torques caused by the misalignment. This accomodation will correct the misalignment. (This is the basis of an entirely logical approach implemented for Ref. 5 and shown in Figure 8). However the 1 in the upper left corner will cause little modification to the reach motion unless forces coming back in reach get to high values, such as when the peg bottoms in the hole, when motion will stop. (In this and all other examples, the quantities in the matrices, except zeros, are representative only).

The basic structure behind all the examples is a vector data and logic loop with characteristics which can be varied by changing the casuality or the matrices inserted in the feedback loop. One may program the arm to act as a damper or as an accomodating linkage, depending on whether the arm is a force source or a motion source. The necessary calculations follow the same pattern and may be programmed into a special purpose computer.

Figure 9 shows the block diagram of such a system, including all the relevant dynamics. The control feedback matrices K_T and K_{TD} are chosen to diagonalize the arm-load dynamics in hand-oriented coordinates. Matrix K_E represents the elasticity of the wrist force sensors and the environment, while matrix K_F is the admittance matrix in Figure 6. A simulation of this system for a two degree of freedom planar arm whose endpoint collides with and follows along a circular barrier is shown in Figure 10. The accomodation loop provides stable behavior and large contact forces do not build up. (θ_R in Figure 9 is the θ -space image of the boundary, represented in θ-space for purposes of illustration only).

This kind of system organization should be viewed as part of a hierarchy whose top level contains the ability to process some descriptive part information and assembly sequence instructions together with a judicious mix of discrete binary tests and accomodation strategies. Below the command level are routines which calculate the necessary feedback matrices, which are then fed down to execution routines which drive the arm using Resolved Motion Rate Control logic and read the discrete and continuous sensors.

4. Sensor Systems

A first laboratory wrist force sensor system has been built (Figure 11) and used in conjunction with the work in ref. 5, where several accomodation-type tasks were performed. This sensor employs a compliant structure instrumented with solid state strain gauges to generate three forces and three torques in hand-oriented coordinates. Figure 12 defines the geometry for resolving the forces and torques about the tip of the large vertical shaft. The necessary computations are given by

$$\begin{bmatrix} 0 & -1 & 0 & -1 & 0 & 0 & 0 & 0 \\ -1 & 0 & -1 & 0 & 0 & 0 & 0 & 0 \\ 0 & 0 & 0 & 0 & -1 & -1 & -1 & -1 \\ -b & +b & +b & -b & 0 & 0 & 0 & 0 \\ -a & 0 & -a & 0 & 0 & -b & 0 & +b \\ 0 & +a & 0 & +a & +b & 0 & -b & 0 \end{bmatrix} \begin{bmatrix} f_{y1} \\ f_{x2} \\ f_{y3} \\ f_{x4} \\ f_{z1} \\ f_{z2} \\ f_{z3} \\ f_{z4} \end{bmatrix} = \begin{bmatrix} F_x \\ F_y \\ F_z \\ M_x \\ M_y \\ M_z \end{bmatrix}$$

where F_x, F_y, F_z, M_x, M_y, M_z are the forces and moments exerted on the shaft and f_{y1}, f_{x2}, f_{y3}, f_{x4}, f_{z1}, f_{z2}, f_{z3}, f_{z4} are the forces measured by the strain gauges, b is the distance from the center line of the feeler shaft to the ends of the compliant bars, and a is the distance out along the feeler shaft to the point where the forces and moments are assumed to be applied.

5. Conclusions

Progress so far indicates that wrist force sensor arrays and accomodation loops are valid concepts which can accomplish assembly tasks involving close fits without requiring precise a priori information concerning part location or geometry. Subsequent developments will include discrete logic tests. Among the additional problems which need to be solved are how to describe assembly operations and how to teach them to an automatic assembler. Providing a structure like Figure 5 and supplying it with appropriate feedback matrices seems like a promising concept.

REFERENCES

[1] Laning, J.H., and Lynde, D., "An Executive Program for Computer-Aided Mechanical Design," MIT Draper Lab Report E-2639, February, 1972.

[2] Barber, D.J., "MANTRAN: A Symbolic Language for Supervisory Control of an Intelligent Remote Manipulator," MIT SM Thesis, Mech. Eng. Dept., 1967.

[3] Hill, J.W., and Bliss, J.C., "Tactile Perception Studies Related to Teleoperator Systems," Stanford Research Institute, Final Report NASA Contract No. NAS2-5409, June 1971.

[4] Whitney, D.E., "The Mathematics of Coordinated Control of Prosthetic Arms and Remote Manipulators," ASME Journal of Dynamic Systems, Measurement and Control, December, 1972.

[5] Groome, R.C., Jr., "Force Feedback Steering of a Teleoperator System," MIT Draper Laboratory Report T-575, August, 1972.

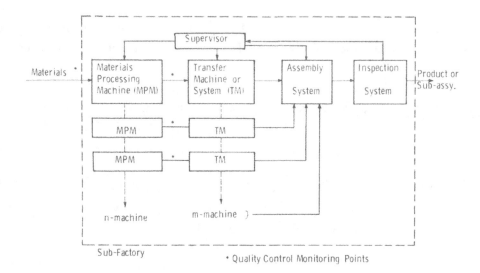

Fig. 1

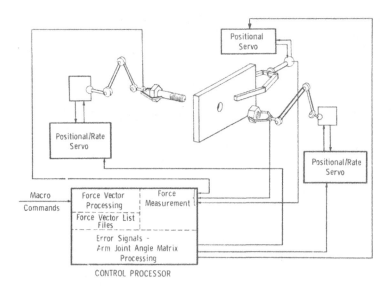

Fig. 2

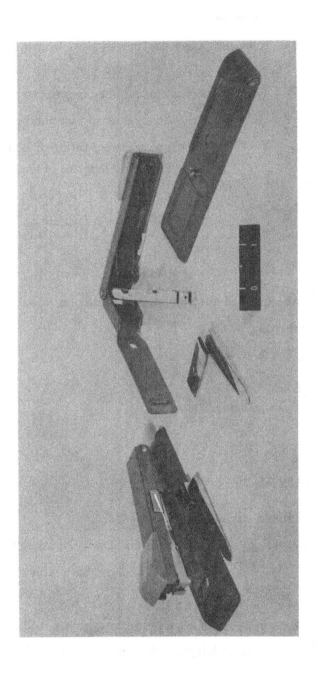

Fig. 3

Assembly description of a stapler

       Notation

1. Location

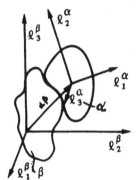

$\alpha, \beta,$ = Represents number assigned to parts

$\ell\alpha - a$ = Distance order a in part number α

$F^{\alpha\beta}$ = Vector of relative position of body α with respect to β. Expressed in terms of the coordinates of β

$$A^{\alpha\beta} = \left\{ a_{ij}^{\alpha\beta} \right\} = \left\{ e_i^{\alpha} \cdot e_j^{\beta} \right\} = \left\{ \cos\left(e_i^{\alpha}, e_j^{\beta} \right) \right\}$$

Matrix of relative rotations

2. Path

– $F^{\alpha\beta}$ and $A^{\alpha\beta}$ are expressed as functions of a parameter λ, such that $\lambda \geqslant 0$ and $\lambda = 0$ gives the final position

$$- \delta(\lambda) = \begin{cases} = 0 \text{ for } \lambda \neq 0 \\ 1 \text{ for } \lambda = 0 \end{cases}$$

– Θ Parameter that can take any value

3. Sequence

    For step C, steps A and B are needed

    For step C, steps A or B are needed

$(\alpha - \beta - \dots) - (\omega - \epsilon - \dots)$ – Represents the activity of assembling subassembly $(\omega - \epsilon - \dots)$ to subassembly $(\chi - \beta - \dots)$.

Note   The description of the path given only expresses it in the neighborhood of the final position. For describing the whole path we have to choose a particular sequence.

                    Figure 3 (Continued)

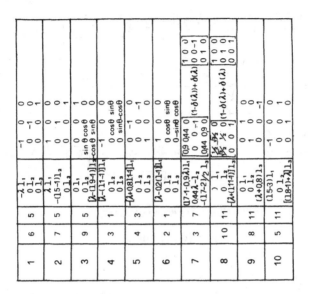

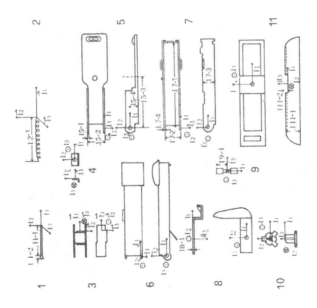

Fig. 3 (continued)

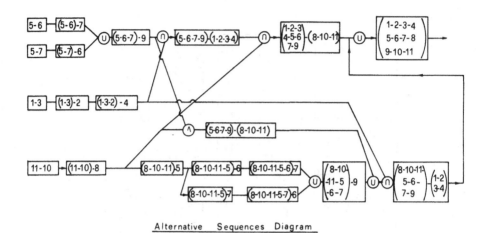

Alternative   Sequences   Diagram

Fig. 3 (continued)

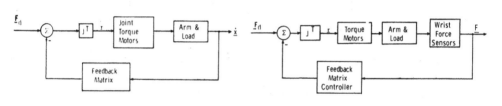

Fig. 4   Fig. 5

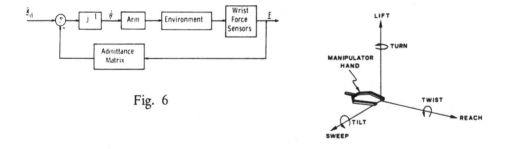

Fig. 6

Fig. 7

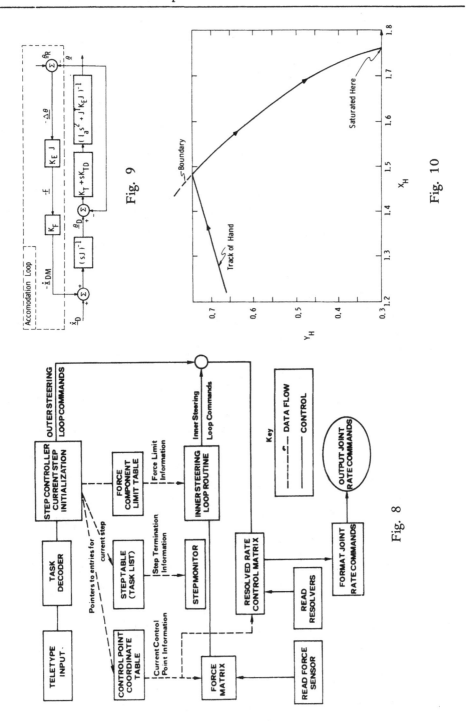

Fig. 9

Fig. 10

Fig. 8

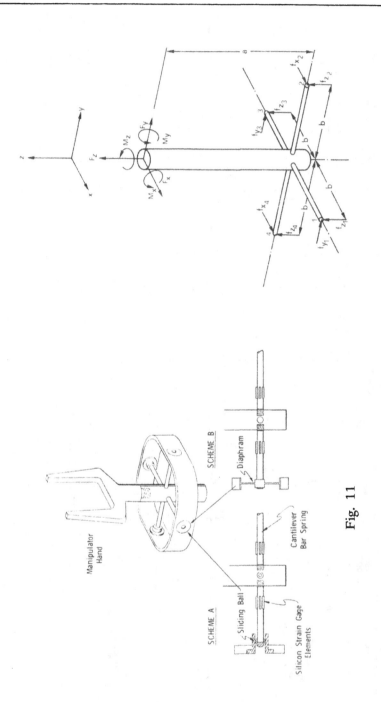

Fig. 12

Fig. 11

ARTICULATED MANIPULATORS
AND INTEGRATION OF CONTROL SOFTWARE

Kohei SATO, Chief of Automatic Control Division,
Electrotechnical Laboratory, Chiyoda-ku, Tokyo, Japan

Kenji OKAMOTO, Systems and Control Section
Hirochika INOUE, Systems and Control Section
Kunikatsu TAKASE, Systems and Control Section

(*)

Summary

  This paper describes the design of articulated manipulators and their control software. A prototype model is an articulated hydraulic manipulator. An advanced one is driven by clutchtendon servo-mechanisms, and the servoing is performed by the software so as to control not only the position but also the speed and the torque. In order to provide a better manipulator system, it is necessary to integrate various control software, ranging from primitive subroutines to supervisory program. This paper exemplifies an interactive operating system and a task-improving supervisor that supports efficient accumulation and modification of the robot's experiences.

(*) All figures quoted in the text are at the end of the lecture.

1. Introduction

Manipulation is one of the important aspects of robotics. Research on general manipulation requires an elaborate machine which performs such tasks as picking up, moving and setting down an object, because it is difficult to simulate these complex manipulation processes. In such a process, the position and the posture of a robot's hand should agree with those of the object with the desired arm configuration. Therefore, a manipulator should have many degrees of freedom of motion and be controlled by a computer, so as to provide general purpose capacity. As regards each servomechanism of the manipulator, not only the position but also the torque and the speed should be controllable, especially in performing dexterous and dynamic actions.

On the other hand, the more elaborate a machine becomes, the more difficult is its control. The integration of control software is a key to providing a better manipulator system. The lowest level software consists of many primitive subroutines which play a role just like the instructions of a computer. For efficient operation in programming, editing and executing a task, it is desirable to prepare an interactive operating system that facilitates better man-robot communication. Besides, efficient accumulation and modification of the robot's experiences will greatly contribute to improve the flexibility of the control software.

This paper describes the hardware of the articulated manipulators, the primitive subroutines, an interactive operating system and a task-improving supervisor, all of which have been developed at the Electrotechnical Laboratory, Tokyo.

2. The ETL-Arm and its primitive subroutines

The ETL(*)-Arm, as illustrated in Fig. 1, is an articulated manipulator which is hydraulically controlled and which consists of an arm and a hand with touch sensors. The arm has six degrees of freedom of motion: shoulder rotation, shoulder pivot, elbow pivot, forearm rotation, wrist rotation and wrist pivot. The manipulator has been designed with special attention paid to its driving mechanism and the entire shape of the system for minimum protrusions, so that the motion of the robot may not be obstructed in a large attainable space. As a result, most of the piping for pressurized oil runs through the robot proper, supplying the oil to the joints by means of swivel joints. The hand consists of a pair of fingers, as shown in

---

(*) ETL   Electrotechnical Laboratory

Fig. 2. Each finger has its own small electric servomotor, and its position and gripping force are independently controlled in a continuous mode. Contact with an object is detected by sensors placed on the fingers.

In operating the arm, the required number of pulses is determined by means of the computer, and these pulses are transmitted to the stepping motors which control each degree of freedom. Once the reference values are imparted through the output voltage of the potentiometers connected to the stepping motors, all axes may be controlled by hydraulic servomechanisms.

Primitive subroutines of a manipulator control system play a role just like the instructions of a computer. They perform the coordinate transformation between the robot hand coordinates and servo angles, and control the motion of the hand. For the robot hand coordinates, an orthogonal coordinate system being fixed in the work space is used. The position of the hand is represented by the coordinates of the finger tip (X, Y, Z), and its posture is represented by the Euler angles (α, β, γ). Servo angles are shown in Fig. 3.

The following are the principal subroutines and commands which have been developed:

MOVE20 (T, X, Y, Z,α , β , γ): This subroutine calculates the six servo angles ($\theta_1, \theta_2, \ldots \theta_6$) corresponding to the position (X, Y, Z) and the posture($\alpha, \beta \gamma$),

in the robot hand coordinates, and moves the arm to the desired state. Argument T specifies the data type (relative or absolute) and whether tactile sensors are utilized or not. This primitive consists of three subroutines called TRANS, THETA and PULTAC. TRANS reads the current value of the potentiometers, and makes a transformation from servo angles to robot coordinates. THETA solves for the servo angles (θ_1 , θ_2 , ..., θ_6) from the hand coordinates (X, Y, Z, α , β , γ). When THETA comes across some deadlock situation, this information is returned before calling PULTAC. PULTAC uniformly interpolates the change in θ_i (i = 1---6) , and puts out adequate pulse trains while monitoring the tactile sensors.

MOVE30 (T, X, Y, α , β, γ , $\Delta\alpha$, $\Delta\beta$): This is an expansion of MOVE20, in which the solution is derived with some tolerances ($\Delta\alpha, \Delta\beta$) permitted to α, β in the calculation of the transformation from robot coordinates to servo angular positions.

MOVE32 (T, X, Y, Z, XX, YY, ZZ, ρ_1 , ρ_2): In this primitive, the solution is obtained by the gradient method. XX, YY, ZZ designate the target value of the elbow position. Both the finger tip position and the elbow position are calculated within their allowance denoted by ρ_1 and ρ_2 , respectively.

MOVE00 (T, i, θ_i): This is the simplest primitive, which moves the motion of i-th servo angle by θ_i .

MOVE10 (T, θ_1 , θ_2 , θ_3 , θ_4 , θ_5 , θ_6 ,   ): An expansion of MOVE00.

INITIAL: Matches the pulse motor angles to the angle of each axis, in order to avoid the initial hazard on applying the oil pressure.

RESET: Returns the configuration of the whole arm to its initial state

CONV20: Measures the servo angles, calculates the robot coordinates of the finger tip, and then prints out by line printer.

CONV21: Calculates and saves the information about the current state of the manipulator.

OPEN, CLOSE: Opens and closes the fingers.

3. Advanced model of an articulated manipulator

In order to perform a partially constrained motion, for example, securely stacking objects, causing it to slide on a table, putting a peg into a hole and turning a crank, the manipulator must drive the object while conforming to the constraint to which the object is subject. Generally, in motions of this type, torque control plays an important role. Moreover, in work in which impulsive movements are required, as, for example, in the action of striking a drum or of driving in nails, the speed of the manipulator also needs to be controlled. The manipulator described in this section has been developed for the purpose of study of such applications as mentioned above.

Fig. 4 provides an explanatory diagram of the servomechanism. The torque for driving the output shaft is determined by the output torque of the magnetic powder clutches and the magnetic powder brakes. Between the transmission torque τ of the magnetic powder clutch (brake) used in this servomechanism and the excitation current i, there exists a proportional relationship as indicated by $\tau = k \cdot i$, and shown in Fig. 5. Accordingly, by controlling the current given to the clutches by means of a computer, the transmission torque is controlled. Servoing is performed by the program that determines the torques of the clutches and the brakes. In the servoing program, the values of the angle and the angular speed detected by the potentiometer and the tachogenerator are read into the computer through an A/D converter, and then, after calculating the clutch torques and the brake torques, control signals are given to each clutch and brake. This servoing cycle is iterated in an adequate sampling period that is much less than the response time of the mechanism. Thus, not only the position but also the speed and the torque of this servomechanism are easily controlled by the software.

Fig. 6 presents an articulated manipulator which is driven by utilizing the abovementioned servomechanism. The design of the arm includes seven degrees of freedom, respectively, shoulder rotation, shoulder pivot, upper-arm rotation, elbow pivot, forearm rotation, wrist rotation and wrist pivot. Besides these, two freedoms for driving the fingers are available. To make the arm light in weight, the servomechanisms are arranged inside the robot's body, and each torque is transmitted for each freedom by making use of stainless steel wires. For each freedom, the arm is always stretched in mutually opposite directions by means of a pair of counteracting wires, so that the wires do not slacken, and backlash may be forestalled. Moreover, any fluctuation of the total length of the wire with changing arm configurations are always absorbed. This manipulator permits the control system to be freely altered according to the program, thereby making it possible to control the position, speed and torque. Accordingly, this manipulator is able to carry out the partially constrained motions and the dynamic actions which could not be performed by the conventional manipulators of position-control type.

4. Interactive operating system of the manipulator

The primitives, in controlling the manipulator, play a role just like instructions of a computer. In order to make effective use of primitives, a supervisory software is necessary. The interactive operating system described in this section utilizes the primitives in the form of conversation through a typewriter, so that the operations run by use of the manipulator may be carried out on-line. When so complex a machine as an articulated manipulator is directly controlled by a typewriter, some miscalculation regarding the manner of using the primitives or any small mistyping or similar error is act to lead to a serious failure. It is, therefore, essential to design a system of this type on the concept of multiple protection for its failsafe operation, so that it may be used care free. This system comprises the following three programs – a command interpreter, an editor and a task executor.

Command interpreter: This program interprets the command given by the typewriter, demands the data to be provided, if necessary; then it selects the desired primitives, and executes it. For the prevention of the operator's careless error, the computer provides guidance of the operation by printing the message. The command given by the operator is checked for possible error as much as possible, so that the next demand may be issued only after a correct command is received. Besides the commands described in section 2.2, those concerned with preparation of experiments, logging experimental data, filing of a task program and so on, are also acceptable.

Editor: This program is used to perform on-line programming and editing of the task to be executed by the manipulator. The composing units of the task are represented as (i, C, L, J) in the computer. "C" denotes the code of the primitive, and "L" indicates the label attached to the data. "J" designates the pointer showing the location of the next element. The data table lists the labels and data, and tells which ones are corresponding with the forms for which primitive. A task is represented by means of linear lists of elements chained by the use of J. In the editor, commands for performing the addition, replacement, insertion or deletion of the primitive-code, the addition, replacement or deletion of data, and the printing of the task program and data are held in readiness. Furthermore, in programming a task, such primitives as unconditional branch, conditional branch, call, return, and so on, are available.

Task executor: A task interactively programmed by using the editor may be immediately executed by the task executor. In the task executor, the task element is traced by making use of i and J. By means of "i", C(i) is read out and decoded. When it requires some data, the data are accessed by means of L(i). After making sure that the forms of data agree with what C(i) demands, the data are delivered to the primitive to perform the action. The next element may be known from J(i). In this way, the task is traced and carried out in successive steps.

5. Task improving supervisor

Efficient integration and modification of the task program will greatly contribute to improve the flexibility of the control software. In order to improve the ability of the operating system, a task improving supervisor has been developed which allows a robot to accumulate and take advantage of its experiences. All the tasks used in this system are represented by the symbols, and so the user begins with defining some sorts of task elements. This system has various basic functions such as to define, connect, shift, linear combination, scale change, edition and so on. For example, function DEFINE defines a task element "GOX" in the expression of GOX = = MOVE20 (T,100,0,0,0,0), which corresponds to a concrete relative MOVE20 (T,100,0,0,0,0). Function COMBIN defines a new task element "SLASH" by a linear combination of GOX and GOY; COMBIN SLASH = GOX + 2 ·GOY. Function SEQUENS makes a task with sequentially chained symbols; SEQUENS STAIR = GOX! GOY! GOX! GOY.

By means of these basic functions, the symbols which are already registered may be available as materials to produce new tasks, which are in turn usable to create more complicated tasks. Repeating these procedures, tasks that reside in the system become more and more sophisticated in quality and quantity. This means that a robot gradually acquires wider experiences, and so new tasks can be effectively produced. For example, suppose the tasks $S0 = S1!$,A2! A3! S4! A5, $S1 = A11! A12! A13, S4 = A41! S42! A43$, and $S42 = A421! A422$, where letter A means task element. Then, if the task Sm is given in the expression of $Sm = A0! S0$, the task Sm is interpreted as equivalent to the following action sequence. $Sm = A0! A11! A12! A13! A2!$ A3! A41! A421! A422! A43! A5! .

Furthermore, in this system, some general functions are available for the automatic creation and improvement of a task. The general function TRNSFR makes up a list of the steps taken by which an object is transferred while avoiding certain obstacles by means of tactile and force sensors. Another general function SMOOTH refines a list of rugged steps into smooth one. Fig. 7-1 shows such a process, in which the steps for bringing the object from the starting point to final point are built up. First, the arm tries to bring the object, move it toward the final point. If some obstacles stand in the way, however TRNSFR provides a route to avoid the obstacles by devising a proper path along which to approach the target. Fig. 7-6 shows the task created by TRNSFR in this manner. SMOOTH rearranges this sequence of steps into smooth one as shown in Fig. 7-7.

6. Conclusions

In this paper, the hardware of articulated manipulators, the primitive subroutines, an interactive operating system and a task improving supervisor have been described.

The ETL-Arm is a prototype model of an articulated manipulator whose positions are controlled by hydraulic servomechanisms, which are powered by magnetic clutches and brakes of continuous torque output. Besides the usual six degrees of freedom, it has an additional freedom of upper arm rotation. This redundant freedom plays an important role in positioning the hand with the desired arm configuration. Its servoing is performed by the program that determines the torques of the clutches and brakes. Thus, not only the position but also the velocity and the torque of this arm are easily controlled by the software, so as to carry out advanced manipulations such as assembling parts, handling tools and driving a nail into a wall.

The control software of the manipulator includes those of varying levels ranging from the primitives to the supervisory program. The primitives, in controlling the manipulator, play a role just like instructions of a computer. In order to utilize them efficiently, and adequate supervisory software is necessary. For such a purpose, an interactive operating system has been developed, so as to supervise the manipulator through a typewriter. Besides, a task improving supervisor has been developed which allows a robot to accumulate and take advantage of its experiences. The development of such kind of software is interesting as it promises to make a robot have more human-like functions. Then, the integration of the control software will play a key role in providing better manipulator subsystem.

REFERENCES

[1]   Special edition: ETL-ROBOT Mk 1, Bul. of Electrotech. Lab., 35, 3, 1971.

[2]   INOUE H.,"Computer Controlled Bilateral Manipulator," Bull. of the Japan Society of Mechanical Engineers, 14, 69, 1971.

[3]   PAUL R., Modelling,"Trajectory Calculation and Servoing of a Computer Arm," Stanford Univ. Ph.D Thesis, 1972

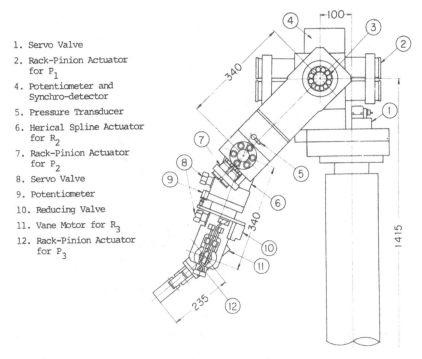

1. Servo Valve
2. Rack-Pinion Actuator for P_1
4. Potentiometer and Synchro-detector
5. Pressure Transducer
6. Herical Spline Actuator for R_2
7. Rack-Pinion Actuator for P_2
8. Servo Valve
9. Potentiometer
10. Reducing Valve
11. Vane Motor for R_3
12. Rack-Pinion Actuator for P_3

Fig. 1   The ETL-Hand

Fig. 2   The gripper of the ETL-Hand.

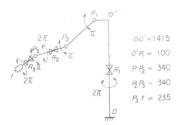

Fig. 3   Freedoms of the ETL-Hand.

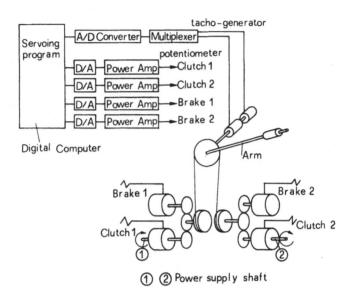

Fig. 4   Explanatory diagram of the clutch-tendon servomechanism.

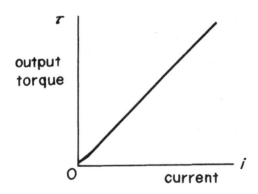

Fig. 5   Characteristics of the magnetic powder clutch.

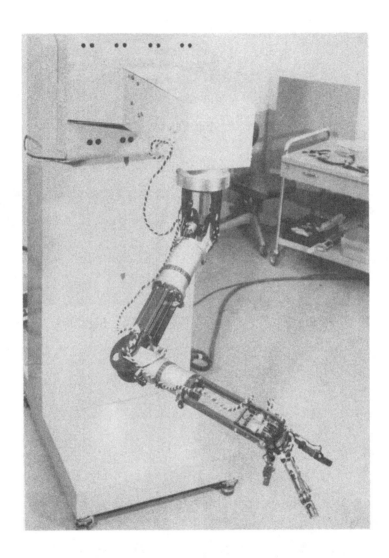

Fig. 6   An advanced model of articulated manipulator.

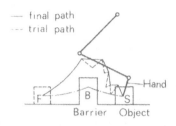

Fig. 7.1

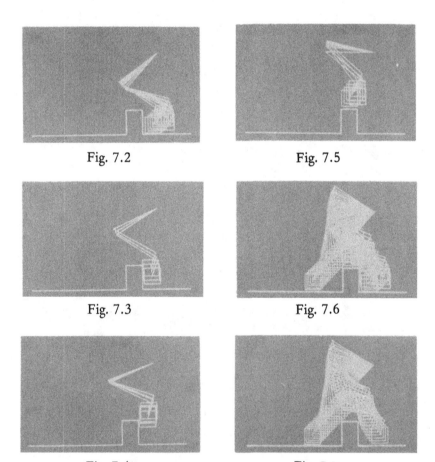

Fig. 7.2

Fig. 7.5

Fig. 7.3

Fig. 7.6

Fig. 7.4

Fig. 7.7

ON MODELING PERFORMANCE OF
OPEN—LOOP MECHANISMS

T.B. SHERIDAN

Department of Mechanical Engineering
Massachusetts Institute of Technology
Cambridge, Massachusetts

(*)

Summary

A technique is presented for modeling the performance of open-loop mechanisms or systems, where performance is measured by success or failure as a function of the selected target for movement along a constrained trajectory and without corrective feedback. A modification of signal detection theory is used, which normalizes dissimilar data and indicates two independent parameters of performance: (1) "discriminability" and (2) "optimality" of the distribution of selected moves. An illustrative application is made to experimental data.

(*) All figures quoted in the text are at the end of the lecture.

Introduction

The control of mechanisms for performing complex sequential tasks such as materials processing, assembly and inspection of manufactured parts, aerospace vehicles and other complex processes can often be characterized as a multilevel process, as indicated in Fig. 1. At the lowest level (the inner loop) position and force variables of the mechanism may be measured continuously or at frequent intervals in time or space (a), then operated upon to drive the mechanism (a′) into continuous conformation with a reference input (b′). The controller at this level is typically a passive electronic or mechanical filter or a computer-based servocontroller.

At an intermediate level (middle loop) a stored program computer is typically used to measure position or force variables (b) of external objects with respect to the mechanism. For example these may be tactile sensors or simple optical ranging or pattern recognition devices. These data are sampled at less frequent intervals than those at (a), and serve together with commands from the human supervisor (c′), as inputs to a computer program which generates a changing reference input to the lower loop (b′).

At the highest level (outer loop) the human supervisor observes with his own eyes directly, or indirectly through aiding instruments such as television, the relation of various objects in the environment (c), including the mechanism, to each other. He acts upon this information, plus his apriori goals or criteria, to command or continually reprogram the computer controller. The man's observations (c) and commands to the computer (c′) are at less frequent intervals than the computer's measurements (b) and reference input changes (b′), which in turn are at less frequent intervals than the feedback (a) interval to the mechanism and the first level control signal (a′).

A different problem in the engineering such as mechanisms or man-machine systems is to evaluate performance, especially with regard to the importance of feedback at various levels as a function of sampling frequency. In effect, a compromise must be found which most appropriately trades errors against sampling frequency of various loops, and permits modeling a variety of systems with few parameters in standardized form.

An approach is suggested which closely resembles the theory of detecting signals in noise (Green and Swets, 1966). Cohen and Ferrell (1969) applied signal detection theory to binary predictive judgement of success or failure in motor skill tasks. Here we modify the theory to allow for multi or continuous level explicit response decision which results in success or failure on each move, and we suggest other changes to make a theory designed for modeling sensory processes suitable for modeling motor processes.

Typically a complete task is specified either as a continuous trajectory or as a sequence of desired discrete states to be achieved, with constraining conditions as functions of position, time and energy. In many cases it is convenient to characterize performance at any point in time by whether it is within a tolerance range (which may be a changing function of position time or energy variables) or whether it is outside this range. Fig. 2 shows schematically two tasks, one defined continuously, the other defined as a sequence of discrete events, both with changing tolerances. "Within" this range means that no significant penalities are incurred and with relative ease it can be brought under control once the loop is closed. "Outside" this range means that the penalty is significant and that bringing the system under control means reverting to an abort mode, a change in normal plan, a return to an earlier part of the sequence, etc. For example a manipulator might accidentally drop an object it was carrying, or confront an unexpected obstacle to its planned path, or find itself in a kinematic singularity or "gimbal lock". A vehicle might run off the road or collide with an obstacle. Thus the sharp nonlinearity in penalty as a function of measured performance variables is characteristic of various real tasks.

We are concerned there with how far along the given continuous path or sequence of discrete stages of the task the mechanism goes in open loop fashion, i.e., before the difference between actual and desired states is measured and corrective efforts are imposed. We will assume that the longer the open loop move the greater the risk of failure, a monotonic relation. We wish to develop both a descriptive model (the statistical description of empirical moves, both the extent of the moves and whether they succeeded or failed) and a normative model (a specification of when the feedback loop should be closed to minimize expected penalty).

The Model

Let x be the target (extent requested) for an open loop move at some given point (a', b', or c') in the system of Fig. 1. If at a, b or c that move is observed to be (or have been during the move) outside the tolerable range, we call it a failure, f. Otherwise it is a success, s. We shall make a plot of the probability density of selecting x, $p(x)$ and the joint probability density of selecting x and observing y to be a failure $p(x,f)$.

Assumptions stated above imply that experimental plots of these quantities will be approximately of the form shown in Fig. 3; the ratio $p(x,f)$ to $p(x)$ increases with x, since the greater x the more the cumulative risk. The joint probability density of selecting x and succeeding is obviously the difference between the two densities,

$$(1) \qquad\qquad p(x,s) = p(x) - p(x,f).$$

By Bayes' theorem the contingent probabilities of x given f or s are

$$(2) \qquad\qquad p(x|f) = \frac{p(x,f)}{p(f)} = \frac{p(x,f)}{\int_0^\infty p(x,f)\,dx}$$

$$(3) \qquad\qquad p(x|s) = \frac{p(x,s)}{p(s)} = \frac{p(x,s)}{\int_0^\infty p(x,s)\,dx}$$

We define cumulative probability functions, i.e., the probability of a selection less than or equal to x,

$$(4) \qquad\qquad P(x|f) = \int_0^x p(x|f)\,dx$$

$$(5) \qquad\qquad P(x|s) = \int_0^x p(x|s)\,dx$$

If a cross plot is made of $P(x|f)$ and $P(x|s)$ as in Fig. 4, a rough equivalent of the receiver operating characteristic (ROC) of signal detection theory results. Though actually it is a distribution of motor responses, we shall call our curve an ROC. This crossplot serves to normalize and standardize the open loop performance characteristic so that two independent parameters of performance are readily apparent. Values of x are scaled monotonically along the crossplot curve.

The first parameter of interest is a measure of "relative discrimination" between success and failure for the particular distribution of x values selected. It is the tendency of the experimental curve toward that ROC "curve" which intersects the upper left hand corner. At this point $P(x|s) = 1$ and $P(x|f) = 0$; therefore it is a point of perfect discrimination, where lesser values of x promise the fullest possible success with no risk of failure, and greater values of x can only be failures. In practice this situation is seldom if ever present, since selecting the greatest x which can be successful usually means a significant probability of failure. This is characterized by an intermediate curve 1 or 2. Diagonal line 0 is where at each x in the distribution $P(x|s) = P(x|f)$. A curve below the diagonal implies a situation where one must experience much failure at smaller values of x to finally attain some successes at greater values. This is contrary to our monotonically increasing risk assumption and thus is a situation we assume to be irrelevant for the present.

The second parameter of interest is the central tendency of $p(x)$ relative to the optimal x. The optimal x is calculated from knowledge of the tendency to failure as a function of x (as estimated from the experimental data), plus the given rewards $R(x)$ for moving as far as x successfully and the given cost $C(x)$ for failure. Thus the expected value of return $E(\mathcal{R})$ over a distribution of moves is

$$E[\mathcal{R}(x)] = \int_0^\infty p(x,s)R(x)\,dx - \int_0^\infty p(x,f)C(x)\,dx \tag{6}$$

However, if it were possible to select consistently a single value of x, one would maximize expected value of return by choosing

$$x_{opt} = \max_x [\mathcal{R}(x)] = \max_x [p(s|x)R(x) - p(f|x)C(x)] \tag{7}$$

The tendency toward the upper left hand corner is the degree of non-overlap of the $p(x|s)$ and $p(x|f)$ distributions. With a given monotonic propensity to fail as a function of x, the percentage overlap of the two distributions can be reduced by selecting over a larger range of x values, insuring success at small x and failure at large x. During learning, such a selection over a large range of x is a good strategy to establish experimentally the dependence of $p(x|s)$ and $p(x|f)$ on x. At later stages presumably the human or computer decision maker attempts to narrow the range of x on a region of near optimal payoff. Necessarily over any narrow range of x $p(x|s)$ and $p(x|f)$ distributions will cease to be displaced relative to each other on the x axis and the ROC curve will approach the diagonal.

There are various means of scaling the tendency to the upper left hand corner. One method used in signal detection theory is to estimate from the data the distance d' between means of distributions $p(x|s)$ and $p(x|f)$ divided by the mean standard deviation. For two Gaussian distributions of equal variance this procedure results in curves for d' = 1 and d' = 2 as indicated in Fig. 4.

Since the slope of the ROC curve

$$(8) \qquad \frac{dP(x|s)}{dP(x|f)} = \frac{d \int_0^x p(x|s)dx}{d \int_0^x p(x|f)dx} = \frac{p(x|s)}{p(x|f)} = \frac{p(s|x)}{p(f|x)} \frac{p(f)}{p(s)}$$

and thus for any given x in a distribution

$$(9) \qquad \frac{p(s|x)}{p(f|x)} = \frac{(\text{slope of ROC})p(s)}{p(f)}$$

From this one might assert that if a single value of x were chosen consistently the return would approximate[3]

$$(10) \qquad \Re(x) \cong (\text{slope of ROC})p(s)R(x) - p(f)C(x).$$

x_{opt} can be found by trial and error by use of equation 10. Measures of the central tendency of x are the average x or, alternatively, the median x. The difference between x_{med} and x_{opt}, if the data warrant a unique specification of the latter, can then be specified. When operating over a narrow range of x where $p(x,s)$ and $p(xf)$ are constant and the ROC has unity slope, (10) reduces to the condition for x_{opt} that

$$(11) \qquad \Re(x) = (\text{constant})R(x) - (\text{constant})C(x)$$

and the choice of x depends only on the relative magnitude and nonlinearity of R and C. The difference between x_{med} and x_{opt}, is a measure of conservatism $(x_{med} < x_{opt})$ or risk $(x_{med} > x_{opt})$ independent of the degree of discriminability.

Example of Application to Experimental Data

In order to illustrate the method, data were taken from two human subjects performing a simple open loop manipulation task which corresponds to the first task of Fig. 2, but with fixed tolerance. For convenience the "mechanism" employed was a human arm and hand holding a simple pen. In this task a subject observed the task, then with eyes closed moved the pen from a starting point toward the right, trying to keep the pen within two lines spaced 0.2 inches apart. The payoff or return function ($\Re$) was

$$(12) \qquad \Re = (x)_{\text{if success}} -(3)_{\text{if fail}} , \quad x \text{ in inches}$$

The subjects were explained the return function and instructed to aggregate the greatest score possible. Each subject had 41 "learning" trials, then 164 "test" trials. On all trials he could obseve what happened on the last trial before going on to the next. The temporal pace was self-determined.

Experimental data for the test trials are plotted in the form of histograms, for movement x in 0.2 inch increments, in Fig. 5. Note the big difference between subjects, subject GF being conservative with small x and small p(f). Subject PS risks large x and large p(f).

Fig. 6 shows the ROC curves approximated from the test data of Fig. 5. Fig. 7 shows the ROC curve drawn for the learning trials of one subject derived by an alternative scheme, namely the point by point plot of sequential values of x. Notice that as predicted the tendency to generate discriminability (d') is greater for the learning trials than for the variability test trials. This is because both absolute range and variability were greater than for the test trials. Notice especially that both subjects' data result finally in approximately the same ROC curves (Fig. 6) even though the raw histograms were quite different. Finally notice that the median responses of both subjects, lie in a region of the ROC where R(x) and slope multiply to keep R(x) relatively constant. This lack of a sharply defined optimum characterizes many tasks.

Conclusions

In designing and controling mechanisms for multidegree of freedom tasks it is imprtant to evaluate performance of open loop moves. If each of a set of responses, intended for open loop movement of extent x, can be characterized by observation into binary performance categories of success and failure, and if a given return function specifies rewards for successful move to x and cost for failure at or before x, and if probability of failure increases monotonically with x, then a performance model analogous to the receiver operating characteristic (ROC) of signal detection theory is useful.

A cross plot of cumulative probabilities of x, given success and of x, given failure, reveals two independent performance parameters. The first parameter is the "relative discriminability" of the set of moves, the degree to which moves of extent less than some value are surely successes and moves greater than some values are surely failures. This index is increased as the range of x of the set of moves increases, and it is decreased by random "noise" or uncontrollability in the mechanism.

The second parameter is the central tendency of selecting or targeting x, relative to the optimal, as determined from the given return function and the empirically demonstrated probabilities of success and failure as a function of x. It is directly a measure of risk vs conservatism in selecting or targeting moves.

An application to data for open loop arm movements by two human subjects illustrates how rather dissimilar distributions yield similar ROC plots. However, the data did not provide sufficiently sharp definition of the "optimal" move distance to assert that the central tendencies tended toward risk or conservatism.

Further development of the analysis and its application are in order.

REFERENCES

[1]   Green, David M and John A. Swets, Signals Detection Theory and Psychophysics, John Wiley and Sons, Inc., New York, 1966.

[2]   Cohen, Harry S. and William R. Ferrell, "Human Operator Decision-Making in Manual Control", IEEE Transactions on Man-Machine Systems, Vol. MMS-10, No. 2, June 1969, pp. 41-47.

[3]   Clearly the abiltiy to selct consistly a value of x and ability to avoid "failure" are interrelated in practical cases, so that the two situations are not directly comparable. The author is grateful to Prof. W.R. Ferrell of University of Arizona for comments on this and related points of the paper.

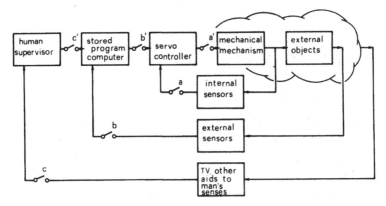

Fig. 1   Manipulator mechanism with multi-level control.

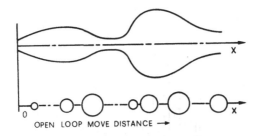

OPEN LOOP MOVE DISTANCE →

Fig. 2   Schematic representation of two open loop task to move along line x, yet stay within boundaries. The upper task is a continuous movement; the lower task is a sequence of discrete movements.

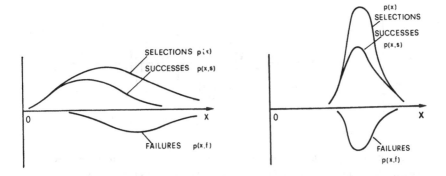

Fig. 3   Typical plots of p(x), p(x,s), p(x,f). The left hand plots represent a case of relatively high discriminability. The right hand plots represent a case of low discriminability, as might occur after a control system settled down in an operating range near the median of p(x) of the left hand distribution.

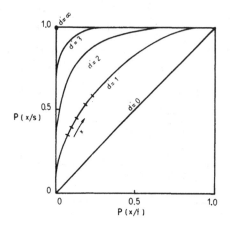

Fig. 4   Cross plots analogous to the receiver operating characteristics
of signal detection theory.

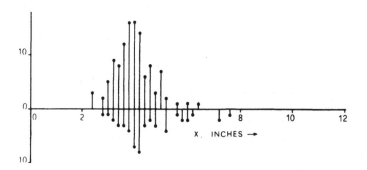

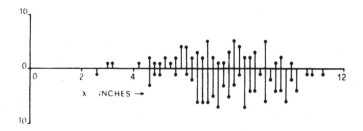

Fig. 5   Histograms of data for two human subjects performing continuous open loop arm movements within a
0.2 inch tolerance zone where the return function = (x/inches) if succed^{-3} is fail.

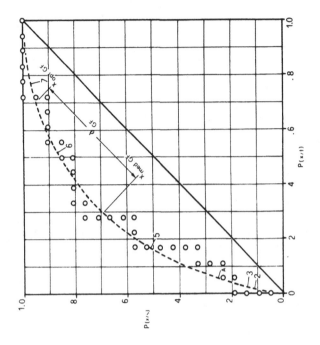

Fig. 7   ROC type crossplot of the learning trials of subject GF, il-
lustrating the method of plotting where each successive x
response is treated as a data point. Notice the relatively
larger d' (higher discriminability) here than in Fig. 6.
Scale along ROC curve is inches.

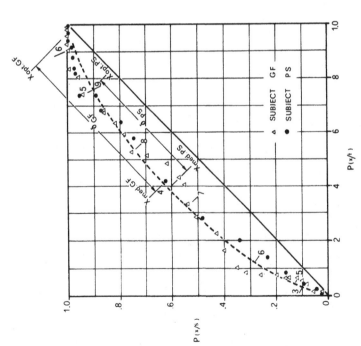

Fig. 6   ROC type crossplots of the test data of Fig. 5. Median and
optimal values are indicated for each subject separately. In
each case the deviation of median from optimal (estimated
from condition that slope of ROC $= 3p(f)/xp(s)$ shows
subjects to be conservative. Numbers above line indicate
inch scale for GF, numbers below line for PS.

DYNAMICS AND CONTROL OF
ANTHROPOMORPHIC ACTIVE MECHANISMS

Miomir VUKOBRATOVIC Ph.D., D.Sc.
Institute Mihailo Pupin, Belgrade, Yugoslavia

(*)

Summary

The study of artificial gait dynamics represents a new class of tasks in the mechanics. The paper describes the method of setting automatically the mathematical models of anthropomorphic mechanisms of arbitrary complexity. The bases how to construct the mathematical models, of artificial biped gait have been presented. It has thus been made possible to synthesize the control that incorporates in itself the automatic maintenance of dynamic equilibrium, as well. The relations between the global indices of the living system control structure and the adopted particular multi-level control system of legged mechanisms artificial motion have been given.

Appropriate specific regulating schemes for maintaining a stable gait under the conditions of system perturbations have been elaborated. In that sense the specificities of the control philosophy of these systems with respect to the classes of regulating processes known so far, have been pointed-out.

(*) All figures quoted in the text are at the end of the lecture.

1. Introductory Considerations

Two directions in solving the control tasks of the complex dynamic systems have mainly been separated so far. One was based on the theory of optimal systems, and the other one did not have any systematization, or rather, it did not have the theoretical background. It has been developed just because of non-efficiency, to be more precise, weakness of the brilliant optimization theory on which an appreciable number of books has been written, while the number of real objects, (processes) whose control has been solved on its fundamentals was just of an inverse proportion. As a result of such a situation the unsolved control tasks have been accumulating in the control theory, for which reason it has been initiated to solve them without some definite control philosophy. Concurrently, a so-called class of near optimal tasks has developed, what has meant in certain way a compromise between the stated task and the possibility offered by the theory. Moreover, a still wider class of complex tasks has been treated based on satisfactory solutions that have resulted from experiments, heuristics, deprived of any regular control concept.

Namely, the basic obstacle in solving control tasks is their multi-dimensionality. The problem of high dimensionality has been solved so far by simplifying the tasks, that is, by reduction to the problem of lower dimensionality. In doing this, may problems have been reduced to linear mathematical models and then attempts made to decompose them based on recognizing the so-called weak connections within the system. (In general, the probelm of high dimensionality was solved in this way). But, be it reminded of the fact that the problem of complex system control has been solved so far only based on linear models by decomposition, which as already known, introduces new, additional interactions to be determined by iterative procedure and optimization technique. Only the statical regimes have been considered then. Dynamic regimes practically remained out of the reach of the apparatus mentioned. In spite of this, the solution found of approximation loses its value at the moment the system operating conditions change, that is, its operating regime, either the change in parameters, or the presence of perturbations are concerned. To expect to be able to master processes in real time in this way (taking not into account the task idealization) would represent in most cases a scientific and engineering utopia.

Keeping in mind such a situation, an attempt has been made to give to these so-called satisfactory solutions some control concept as a basis, that would make it possible to solve various complex tasks by a common method. The procedure proposed is based on hierarchical structure with the following levels [1]:

　　　　　　　　— level of nominal dynamics (algorithmic level)

　　　　　　　　— level of adaptation (level of perturbed regimes).

The significant novelty is the introduction and synthesis of the nominal dynamic regimes. The paragraphs to follow will deal with the synthesis of the preceding control levels.

2. Synthesis of Nominal Dynamics-Method of Prescribed Synergy

　　　　　　　　There exists a wide range of control problems where we are interested only in some definite dynamic states. In such cases it is possible to introduce the term of prescribed synergy to the entire system or to some of its parts. If we restrain ourselves to the most interesting case of partial synergy prescription we come to a specific procedure for us of system dimensionality reduction. It is quite obvious that here neither some system simplification nor its mathematical linearization are concerned. Here only dynamics is prescribed to a system part, and from remainig "open" dynamics of the system it is required to bring the process to equilibrium based on task specificity and corresponding conditions (dynamic connections). The dynamics thus computed can be named the compensating synergy of the dynamic process in genral. By prescribing various synergies we can get an adequate set of compensation synergies and so substitute all dynamic possibilities of the system by those dynamic forms being of interest for the case considered.

　　　　　　　　The mathematical formulation of system nominal dynamics synthesis (prescribed and computed — compensating) could be carried out in the following way [2].

　　　　　　　　Let $\{M\}$ be vector of driving generalized forces of an arbitrary type dynamic process, and$\{\Xi\}$the vector of dynamic state of appropriate process.

　　　　　　　　Then the arbitrary dynamic process can be represented by the following set of nonlinear differential equations:

$$\{M\} = [A]\{\ddot{\Xi}\} + [B]\{\dot{\Xi}^2\} + [C]\{\dot{\Xi}\dot{\Xi}\} + \{G\} \tag{1}$$

The coefficients of matrices $[A]$, $[B]$ and $[C]$, and columns $\{G\}$, as well, depend on the adopted generalized coordinates of dynamic process. This means, eq. (1) represents any dynamic process of a mechanical, electrical, and even biological or economical system, under the provision that it is necessary to introduce certain analogies between the elementary parameters, as for example:

　　　　　　　　— generalized inertia and mass,

— generalized damping,
— generalized stiffness,
— generalized active (driving) forces.

Analogously to the tasks in mechanics, the following three cases can be defined corresponding to the first, second and combined type of the problem:

(1) In the simplest case, motion (in mechanical sense), i.e. dynamics of the complete system (in general sense) is known. Then the resulting set of equations for unknown generalized forces (in general case) becomes a trivial set of algebraic equations wherefrom it is possible to compute the control actions of dynamic process.

(2) But, to determine the dynamics (motion) of a prescribed control action (generalized forces of dynamic process), the second derivative, that is, generalized inertia (acceleration) of the system (1) should be explicitly expressed as:

$$(2) \qquad \{\ddot{\Xi}\} = [A]^{-1}\{M\} - [B]\{\dot{\Xi}^2\} - [C]\{\ddot{\Xi}\Xi\} + \{G\}$$

The resulting second-order differential equations system can be solved by numerical methods assuming the initial conditions $\{\Xi\}$ and $\{\dot{\Xi}\}$.

(3) In the combined type of dynamic task the driving-control actions (generalized forces), and dynamics of the system (process) particular part as well, are known. Given the known generalized forces and dynamics (motion) denoted with M_o and Ξ_o, respectively, and their unknown parts with M_x and Ξ_x. In compliance with such an assumption, introduce transformation matrices $[P]$ and $[R]$:

$$(3) \qquad \begin{bmatrix} P_o \\ \hline P_x \end{bmatrix}\{M\} = \left\{\frac{M_o}{M_x}\right\} \text{ and } [R_o \mathbin{!} R_x]\left\{\frac{\Xi_o}{\Xi_x}\right\} = \{\Xi\}$$

which lead the basic equation (1) to the form of:

$$(4) \qquad \left\{\frac{M_o}{M_x}\right\} = \begin{bmatrix} P_o \\ \hline P_x \end{bmatrix}[A][R_o \mathbin{!} R_x]\left\{\frac{\ddot{\Xi}_o}{\ddot{\Xi}_x}\right\} + \begin{bmatrix} P_o \\ \hline P_x \end{bmatrix}\left([B]\{\dot{\Xi}^2\} + [C]\{\ddot{\Xi}\Xi\} + \{G\}\right)$$

By multiplying matrix $[A]$ with two transformation matrices the result 2 :

$$(5) \qquad \begin{bmatrix} P_o \\ \hline P_x \end{bmatrix}[A][R_o \ R_x] = \begin{bmatrix} A_{oo} & \mathbin{!} & A_{ox} \\ \hline A_{xo} & \mathbin{!} & A_{xx} \end{bmatrix}$$

where submatrices A_{ox} and A_{xo} are symmetric due to the fact that the sets of known generalized forces and motions (dynamics) are complementary.

The eq. (4) can be rearranged in order that the unknowns M_x and $\ddot{\Xi}_x$ appear explicitly:

$$\left\{\begin{matrix} M_x \\ \ddot{\Xi}_x \end{matrix}\right\} = - \left[\begin{matrix} 0 & \vdots & A_{ox} \\ -I & \vdots & A_{xx} \end{matrix}\right]^{-1} \left(\left[\begin{matrix} -I & \vdots & A_{oo} \\ 0 & \vdots & A_{xo} \end{matrix}\right] \left\{\begin{matrix} M_o \\ \ddot{\Xi}_o \end{matrix}\right\} + \left[\begin{matrix} P_o \\ P_x \end{matrix}\right] \left([B] \{\dot{\Xi}^2\} \right. \right. +$$

$$\left. \left. + [C] \{\dot{\Xi}\dot{\Xi}\} + \{G\} \right) \right)$$

(6)

The system (6) consists of algebraic and differential equations, and can be divided into two subsystems for unknown dynamics (system generalized coordinates) and unknown generalized forces:

$$\{\ddot{\Xi}_x\} = [A_{ox}]^{-1} (\{M_o\} - [A_{oo}]\{\ddot{\Xi}_o\} - [P_o] ([B] \{\dot{\Xi}^2\} +$$

$$+ [C] \{\dot{\Xi}\dot{\Xi}\} + \{G\}))$$

(7)

$$\{M_x\} = [A_{xx}]\{\ddot{\Xi}_x\} + [A_{xo}]\{\ddot{\Xi}_o\} + [P_x] ([B] \{\dot{\Xi}^2\} +$$

$$+ [C] \{\dot{\Xi}\dot{\Xi}\} + \{G\})$$

(8)

So far, we have spoken quite generally about the dynamic systems or processes. In the system synthesis, the three above mentioned types of the task are encountered mainly. Particular processes (systems) by their specificity, can somehow modify the task type, but without any significant change. This will be dealt with when describing particular groups of tasks by applying the suggested method of prescribed synergy. As for the method application the third type of tasks is interesting, in further text we shall consider only it. The system (6), that is, subsystems (7) and (8) represent the mathematical basis for the synthesis of nominal dynamic regimes based on the method of prescribed synergy.

3. Adaptation Level Synthesis

The synthesis of this level relies upon the control scheme illustrated in Fig. 1 in its most general form.

According to this scheme [3], the control in the system perturbed regime can be achieved in two ways. Based on the selected criterion and measured deviations between the real and ideal (nominal) synergy, the correction is performed along a contour (loop) by means of servo-systems, that in certain processes, as it will be recognized from the forthcoming sections, should satisfy particular requirements because of unusual operating conditions. Along some other contour, it is possible to select any of the synergies available being prescribed in advance and memorized in the control system, for the case of greater deviations. Thus the process can be extended by the newly chosen synergy, or gradually brought back to the initial operating regime. In this paragraph we will not consider in more detail the description of the adaptive control level synthesis since it will be dealt with in particular classes of control tasks when certain specificities of this generalized idea so far, will appear.

For the moment the essential suitability of this control concept should be emphasized. It consists of a realistic possibility of controlling the complex dynamic processes in real time. This means, many dynamic processes at the present level of computer capabilities cannot be controlled based on on-line computation of new dynamic states, and consequently of separate correcting control signals. Nevertheless, in some cases the process computer cannot be accepted from the standpoint of its price, dimensions, and others. The suggested control concept degenerates in fact the computational complex into an electronic simulator or programmer of dynamic process synergies. The dynamic process synergies, needed for the system operation are memorized in the electronic simulator and thus certain prediction of conditions under which the system can find itself during its functioning, is realized. By such a control the response time of the control system is reduced to the time needed to measure the process dynamic parameters, identify new synergies (in case of greater deviations), compute relatively simple algebraic relations in accord with the selected control type and time of dynamic process actuator performance. It should be pointed-out that the time needed to execute the mentioned phases of control cycle is far less than the on-line computation of new dynamic states by the use of process computers.

Independent on the way of compensating the perturbed regimes of system operation, still there remains the question of deviation criterion between the real and ideal synergy. It will be treated now when the new control concept is considered generally.

Conditionally, the criteria can be divided into two groups global and local. The global are those that estimate the deviations between the real synergy Ξ and ideal synergy Ξ_o in finite time interval. Local criteria estimate these deviations continuously in time.

In many more delicate processes of compensation, a change in acceleration of particular system parts also appears as a direct result of the change in some generalized load, or generalized forces on the dynamic process. To estimate in this case the efficiency of compensating actions, the criterion should include the extended state vector that incorporates second derivatives, as well, of the dynamic process synergy considered.

In the case of local criterion, the deviation between the real and ideal (nominal) synergy is proposed in the following form:

$$J_i(t) = [c_{oi}(\xi_i(t) - \xi_i^o(t)) + c_{1i}(\dot{\xi}_i(t) - \dot{\xi}_i^o(t)) + c_{2i}(\ddot{\xi}_i(t) - \ddot{\xi}_i^o(t))]^2 \qquad (9)$$

where c_{oi}, c_{1i}, c_{2i} — weighting coefficients

ξ_i^o — coordinates of nominal synergy Ξ^o

ξ_i^1 — coordinates of real synergy Ξ

i — Number of generalized coordinate.

The criteria for each generalized coordinate can be treated as the components of vector J.

$$J = (J_1, J_2, \ldots, J_n) \qquad (10)$$

The task of dynamic process compensating system would be to reduce the value of criterion to a minimum. Thus the objective can be stated to find such compensating generalized forces that would reduce the value of criterion (9) to zero during the time interval τ.

Based on relation between ξ and $\dot{\xi}$ in close time intervals t_1 and $t_2 = t_1 + \tau$

$$\xi_i(t_2) = \xi_i(t_1) + \dot{\xi}_i(t_1) \cdot \tau$$
$$\dot{\xi}_i(t_2) = \dot{\xi}_i(t_1) + \ddot{\xi}_i(t_1) \cdot \tau \qquad (11)$$

it is possible to write the criterion (9) for $t = t_2$ as a function of current time t_1

coordinates. Making this criterion equal zero, $J_i(t_2) = 0$ and solving it for $\xi_i(t_1)$, it is possible to find values of acceleration needed to be communicated to the system in order to reduce the criterion to zero after the time period τ

$$\ddot{\xi}_i(t_1) = [c_{2i}\ddot{\xi}_i^o(t_2) - c_{oi}(\xi_i(t_1) + \dot{\xi}_i(t_1)\cdot\tau - \xi_i^o(t_2) -$$
$$- c_{1i}(\dot{\xi}_i(t_1) - \dot{\xi}_i^o(t_2)]/(c_1\tau + c_2)$$

Since the nominal values of the synergy and its derivatives $\dot{\xi}_i^o$ and $\ddot{\xi}_i^o$ are known, by measuring the new state vector

(13)
$$Y = \left\{ \frac{\{\Xi\}}{\{\dot{\Xi}\}} \right\}$$

it is possible to get from expression (12) the acceleration increment (generalized) inertia) of the system $\{\Delta\ddot{\Xi}\}$, and based on eq. (2) compute the vector of compensating generalized forces as

(14)
$$\{\Delta M\} = [A]\{\Delta\ddot{\Xi}\}$$

In this way such additional compensating generalized forces have been determined by means of which the criterion (10) can be reduced to zero during the interval τ with an accuracy up to small second-order values.

The described control scheme based on specific criterion incorporating in itself equally the second derivative of dynamic system synergy, too, is unavoidable in all processes in which it has to be made possible to track directly the variable load (generalized forces), requiring on its part independent feedbacks for forces (load). Some other deviation criteria will be described when considering particular types of the tasks to which the proposed control concept is to be applied.

4. An Example of Synthesis of Artificial Anthropomorphic Gait

The locomotion activity and the biped gait in particular belong to the class of most automated motions. This fact has been recognized by Bernstein [5] already. The locomotion legged, and the antropomorphic system in particular represent extremely complex dynamic systems both form the aspect of mechanical--structural complexity, and of complex control system. Here the fact should be emphasized that nearly 350 pairs of muscles are available for complete skeletal activity of the man. To control a system with so many degrees of freedom can be considered as absurdity. This fact is even more endangered because the human control mechanism is mostly unknown. It is of interest here to cite a global control philosophy of the human locomotion, stated by Bernstein already many years ago. For that reason it is given in the form of its original block-scheme in Fig. 2.

From this block-scheme it is possible to recognize the block denoting the programmer, in which the biological, natural sysnergies of the human skeletal activity, and of gait in particular, are actually memorized. The gait control concept proposed here, possesses an absolute analogy with the natural control system in view of the statement that the prescribed synergy extends some chance to the control of complex dynamic processes in real time, because the nature a priori solves the control problems in the most reasonable way. For further consideration of these analogies use Figs. 1 and 2.

By following the mentioned global indices of human control organization and taking into account the failures to solve the problem of the artificial and of biped motion in particular by means of any of classical approaches of mechanics and control theory so far the following task has been formulated: to realize and maintain a stable anthropomorphic gait under the conditions in which the subject can find himself needing an additional driving power, or the patient to whom the walking machine should serve as a rehabilitation appliance. In doing this the gait type have been "borrowed" from the man (prescribed synergy), and from his extremely complex control system only the hierarchical structure recognized [6,7].

In other words, biomechanical information on how the man walks have been used, whereby the explanation was not searched as to why the gait type is as it is, and the like. Namely, the problem of artificial gait synthesis belongs to the group of tasks in whose mathematical formulation it is not possible to establish some firm optimization criterion based on which it could be made possible to get dynamic state of steady-state regimes. It is resonable to believe, and even to state that in different locomotor tasks of the man, various criteria satisfy, too. In addition, at least two formal reasons can be formulated not contributing to the attempt of solving the problem at any price in an optimal way:

(a) to have the optimization task of any sense in general, the leg joints must be allowed certain freedom in motion and after that constraints introduced in such a defined mathematical model. Such an approach makes the system non-solvable in mechanical-control sense;

(b) If such difficulties could be even overcome, there is no guarantee that the so-called optimal control would even give anthropomorphic trajectories.

As the proof on how difficult and purposeless is the optimization task of anthropomorphic gait synthesis could serve the work by Chow and Jacobson [8]. They have obtained, based on the criterion of minimum energy, the values of driving torques in the hip- and knee-joints, and the appropriate "optimal" trajectories of the corresponding joints, as well. To make the problem solvable, the authors have simplified the anthropomorphic mechanism considerably. Besides others, the motion problem (gait) has been reduced to saggital plane, the gait stability, that is, dynamic equilibrium problem has been not solved at all, and the problem of unknown dynamic reactions (what makes the real gait problem unsolvable in the classical sense of mechanics) has been bypassed by introducing them as the known values.

The mentioned task of antropomorphic gait synthesis illustrates perhaps at best the class of problems where it is needed to search the so-called satisfactory solutions out of the scope of classical control theory and optimization procedures. In this case it is evident that the satisfactory solutions are all those based on various gait types, called the prescribed synergy of lower extremities in accord with our concept.

When we have exposed general mathematical fundamentals of the level synthesis of nominal dynamic regimes, we have mentioned some three types of the tasks. Then it was stated that for us the most interesting is the combined type of the task where the driving-control actions (generalized forces), and a part of process dynamics are partially known.

When the synthesis of anthropomorphic motion is concerned, the mentioned type of the task gets certain specificities. Namely, the problem of biped locomotion, representing in mechanical sense a variable structure, has one very unpleasant specificity: in spite of the prescribed synergy to one part of the anthropomorphic mechanism, in the latter considered as a whole neither the driving torques nor the accelerations are known. In addition to all this, dynamic reaction forces, being the functions of both the prescribed and unknown, compensating synergy (*), that is, its derivatives, act upon the anthropomorphic system as the external forces. Such an inconvenient indefiniteness should be somewhere interrupted. For this reason in the synergy synthesis (nominal dynamic of artificial gait) the

---

(*) In this task the compensating movements of the body upper part represent the compensating synergy, including also the upper extremities in the general case.

attention has been focussed to solving the problem of such dynamic-control indefiniteness. Therefore the points at which the resulting reaction forces act on the contact surface between the extremity and the ground have been prescribed. Those points have been named the zero-moment points [6,7] and with respect to the system of coordinates related to the momentary positions of these points specific equations of dynamic equilibrium have been formed. By doing this, this unusual task of mechanics was practically solved.

This means, when considering an anthropomorphic configuration, the movements of particular mechanism segments are known, thus giving the gait type, while according to the adopted zero-moment points, the moments round the support points and joints of passive elements (*) are equal to zero (Fig. 3). In this case the moments $\{M_0\}$ are zeroes and the system of differential equations for unknown angles (7) is as follows:

$$
\{\ddot{\Xi}_x\} = -[A_{ox}]^{-1} ([A_{oo}]\{\ddot{\Xi}_0\} + [P_0]([B]\{\dot{\Xi}^2\} +
$$
$$
+ [C]\{\dot{\Xi}\dot{\Xi}\} + \{G\})) \tag{15}
$$

The appropriate driving torques of particular active joint $\{M\}$ can be found from the system (1), where the angles have been determined in compliance with eqs. (3), that is,

$$
\{\Xi\} = [R_0 \vdots R_x] \left\{ \frac{\Xi_0}{\Xi_x} \right\} \tag{16}
$$

The unknown angles $\{\Xi_x\}$ in this case describe the motion of compensating parts that is needed for equilibrium round the support point, as well as the free motion of passive elements.

But, with this the mathematical problem of biped gait has been not solved. Still nothing is known about the unknown angles at the step beginning and end. The initial and end states should follow the repeatability conditions that consist of the following: all $\{\Xi\}$ and their derivatives $\{\dot{\Xi}\}$ at the beginning and end of the step period should be equal, i.e.

$$
\{\Xi\}_T = \{\Xi\}_0 \quad \text{and} \quad \{\dot{\Xi}\}_T = \{\dot{\Xi}\}_0
$$

(*) In the anthropomorphic mechanism synthesis, the upper extremities are treated as the system passive elements.

In the known part of angles $\{\Xi\}$, that is, $\{\Xi\}$ these conditions will be satisfied when the data taken on gait correspond to the stationary gait. Thus the repeatability conditions for the unknown (compensating) part,

(17)
$$\left\{ \frac{\Xi_x}{\dot{\Xi}_x} \right\}_T = \left\{ \frac{\Xi_x}{\dot{\Xi}_x} \right\}_0$$

represent the boundary conditions of the problem.

      This problem with boundary conditions can be solved by linearization in the neighborhood of approximate solution by means of successive corrections obtained with the sensitivity matrix [3,4]:

(18)
$$\left\{ \frac{\Xi_x}{\dot{\Xi}_x} \right\}_T + \left\{ \frac{\Delta\Xi_x}{\Delta\dot{\Xi}_x} \right\}_T = \left\{ \frac{\Xi_x}{\dot{\Xi}_x} \right\}_0 + \left\{ \frac{\Delta\Xi_x}{\Delta\dot{\Xi}_x} \right\}_0$$

      The changes in finite conditions can be found by means of sensitivity matrix based on changes in initial conditions, i.e.

(19)
$$\left\{ \frac{\Delta\Xi_x}{\Delta\dot{\Xi}_x} \right\}_T = [\,U\,] \left\{ \frac{\Delta\Xi_x}{\Delta\dot{\Xi}_x} \right\}_0$$

where $[U]$ is the sensitivity matrix

(20)
$$[\,U\,] = \begin{bmatrix} \dfrac{\partial(\Xi_i)}{\partial(\Xi_j)} & \dfrac{\partial(\Xi_i)}{\partial(\dot{\Xi}_j)} \\ \dfrac{\partial(\dot{\Xi}_i)}{\partial(\Xi_j)} & \dfrac{\partial(\dot{\Xi}_i)}{\partial(\dot{\Xi}_j)} \end{bmatrix}$$

Eqs. (18) and (19) give necessary corrections for initial conditions, that is,

(21)
$$\left\{ \frac{\Delta\Xi_x}{\Delta\dot{\Xi}_x} \right\}_0 = (\,[U] - [I])^{-1} \left(\left\{ \frac{\Xi_x}{\dot{\Xi}_x} \right\}_0 - \left\{ \frac{\Xi_x}{\dot{\Xi}_x} \right\}_T \right)$$

Because of the step symmetry, the repeatability conditions (17) can be expressed by conditions for the half step

$$\left\{ \frac{\Xi_x}{\dot\Xi_x} \right\}_{T/2} = [\,W\,] \left\{ \frac{\Xi_x}{\dot\Xi_x} \right\}_0 \tag{22}$$

where [W] is the matrix giving the state relation at the step beginning and its half. Thus, analogously to relation (21) one can write

$$\left\{ \frac{\Delta\Xi_x}{\Delta\dot\Xi_x} \right\}_0 = (\,[U] - [W]\,)^{-1} \left([W]\left\{ \frac{\Xi_x}{\dot\Xi_x} \right\}_0 - \left\{ \frac{\Xi_x}{\dot\Xi_x} \right\}_{T/2} \right) \tag{23}$$

As far as the compensating mechanism in the case of gait perturbed regimes is concerned, there exists a number of specificities due to which the general considerations on process stabilization based on the proposed control concept, should be extended by certain additional considerations. It has to be underlined that expressions (9) − (14) can represent rather different system classes, so that in that section of the paper no details were mentioned that could lead to specific properties related to a definite type of the task. When the stabilization of antrhopomorphic gait is dealt with, the following formulation should be emphasized: the anthropomorphic active mechanism is stable if during the gait, exposed to the effects of perturbations, it returns to the old stationary regime or is re-established on the new one in accordance with the general block-scheme in Fig. 1. Further, the biped gait stabilization has a number of specificities that directly make impossible the classical control of the gait process.

First, one of the basic specificities is that the coordinate φ_0 (Fig. 4) is noncontrollable. The compensation for φ_0 is possible only at the expense of the change in load (driving torques) of other coordinates (φ_i, i = 1,2,...,n). The conclusion can be made that for the purpose of providing stability, it is necessary to have the possibility of controlling the locomotion system driving torques. In view of this in Fig. 1, being of general significance for the suggested control concept, the term robot should be placed into the block provided for object for the concrete type of the task.

The next specificity is that the consideration should include the terms of the so-called internal and external synergy (Fig. 5). Be $\varphi_i(t)$ the coordinates of internal synergy, while $\beta_i(t)$ are coordinates of external synergy determining the system position with respect to the fixed system (positions of zero-moment points).

In the case of perturbations (Fig. 4), even when the internal synergy φ_i (t) is strictly fulfilled, the external synergy can be perturbed. For example, the entire model, as a rigid body, can turn round the support of stance leg and during that the angles β_i change. This means that the reasons for change in external synergy are manifold. Due to any reasons be the synergy β perturbed with respect to its nominal value β^o. The compensating effects of the driving system in this case represent internal forces that change directly the internal synergy φ (t).

Two, in principle different ways of providing process stability, given in Fig. 1, in view of this task are reduced to the following:

(a) starting from real synergy β it is possible to select a new nominal internal synergy φ^o, under the provision the real synergy β corresponds to the ideal synergy β^o;

(b) the internal synergy can be considerably violated in order to approach the external synergy to the initial ideal synergy β^o_i.

As in the general description of the proposed control concept, when a rather general criterion of deviation between the process real synergy and the nominal one was dealt with (9), such a criterion must be applied here by all means. Write this criterion analogously to form (9) as:

(24)
$$J_i \ (t) = [\ c_{oi} \ (\beta_i \ (t) - \beta^o_i \ (t)) + c_{1i} \ (\dot{\beta}_i \ (t) - \dot{\beta}^o_i \ (t)) \ + \\ + \ c_{2i} \ (\ddot{\beta}_i \ (t) - \ddot{\beta}_i(t))]^2$$

Here, i is the number of segments of the anthropomorphic mechanism.

Further, the control procedure develops indentically to that one given for the general case by the expressions (11), (12) and (14). In this case, due to the task specificity in which the external synergy cannot be modified directly by the compensating system, it is necessary to find relation between the angles of external and internal synergy

(25) $\beta = B\varphi + C$ $C = const.$

where B is constant matrix.

From (25) we get the relation

(26) $\ddot{\varphi} = B^{-1} \ddot{\beta}$

Then, from the general expression for compensating effects (14), it is possible, in concrete case, to get relation between the compensating driving torques and the acceleration increments of external synergy as

$$\{\Delta M\} = AB^{-1}\Delta\ddot{\beta} \qquad (27)$$

5. Active Exoskeleton — The First Application of the New Control Concept

For the purpose of applying practically the proposed control concept an active exoskeleton has been realized intended for rehabilitation of paralysed people with sufficiently high lesion of the spinal cord for which reason the muscle activity of lower extremities has been reduced to zero [9,10,11]. Fig. 5 illustrates the last version of the complete exoskeleton that has proved the simplicity of the new concept of algorithmic control. The characteristic of this machine is in its minimum number of degrees of freedom to be controlled, being sufficient for performing the elementary locomotion activities (gait upon level ground and stairs). The anthropomorphic mechanism as a whole is stabilized by means of measuring dynamic reaction forces whereby only "Good" Tendency of perturbations compensation has been provided. The system: exoskeleton — patient is subject to those perturbations during the gait. In this phase of system realization, the incomplete control scheme not caring of adjusting the complete state vector, is compensated with "stabilizing canes" whose only task is to provide the patient with a stable gait during his training, and to give him a feeling of security in the present phase of exoskeleton medical evaluation. Since the mentioned anthropomorphic exoskeleton with the patient is not able of realizing automatically a complete dynamic equilibrium because of understandable reasons, there always exists a deviation with respect to the nominal dynamics and in the system compensating part in particular, due to which the programmed gait will be perturbed. As the result, the characteristic parameters of gait, S (step length) and T (step period) are violated, for which reason the forseen gap between the foot and ground disappears, and the foot strikes on the ground. Therefore the so-called algorithm flexibility has been secured, that is aimed to providing the sequence in performing particular step phases, and the phase ratio between one and the other leg according to a forseen program (prescribed synergy). This problem seriously endangering the concept of prescribed synergy, when it is impossible to realize it sufficiently accurately, has been solved simply by incorporating a parir of switches into each foot; these switches response to the electronic logic in the a.m. sense of providing the sequence of program phases.

The results of experiments have shown that the concept of "time" correction of the synergy imposed to the biped machine lower extremities is fully justifiable.

The next, perhaps most significant property of such a conceived machine is that the process is performed based on memorized synergy placed in the electronic simulator, called the programmer, while the adaptation to he modified operating conditions is done in real time. Within the proposed general control concept, and in this concrete task the computer has been avoided, which would make the practical applicability of the exoskeleton as a rehabilitation appliance questionable.

REFERENCES

[1] Vukobratović, M., Stepanenko, Y., "Mathematical Models of General Anthropomorphic Systems", Mathematical Biosciences, (in press).

[2] Juricić, D., Vukobratović, M., "Mathematical Modelling of Bipedal Walking System", ASME Publication 72-WA/BHF-13.

[3] Vukobratović, M., "How to Control Artificial Anthropomorphic Systems" IEEE Transactions on Systems, man and Cybernetics, (in press).

[4] Vukobratović, M., Stepanenko, Y., "On the Stability of Anthropomorphic Systems", Mathematical Biosciences, 15, 1-37, 1972.

[5] Bernstein, N.A., On the Motion Synthesis (in Russian), Medgiz, 1947.

[6] Vukobratović, M., et al., Final Report to SRS: "Restore the Locomotion Functions to Severely Disabled Persons", 1972, M. Pupin Institute.

[7] Vukobratović, M., et al., "Analysis of Energy Demand Distribution with Anthropomorphic Systems", Proc. of IV Symp. on External Control of Human Extremities, Dubrovnik, 1972.

[8] Chow, C.K., Jacobson D.H., "Studies of Human Locomotion via Optimal Programming", Technical Report No. 617, 1970, Division of Engineering and Applied Physics, Harvard University.

[9] Vukobratović, M., et al., Progress Report to SRS (Social Rehabilitation Service), "Restore the Locomotion Functions to Severely Disabled Persons", No. 1, 1969/70.

[10] Hristić, D., Vukobratović, M., "A New Approach to the Solution of Rehabilitation System for Disabled Persons", Automatika No. 3, Zagreb, 1970.

[11] Vukobratović, M., et al., "Development of Active Exoskeletons", Medical and Biological Engineering, (in press), 1973.

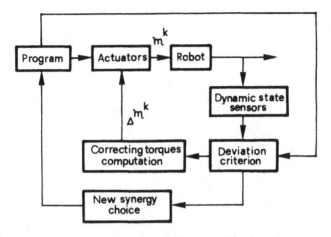

Fig. 1 General stabilization block scheme

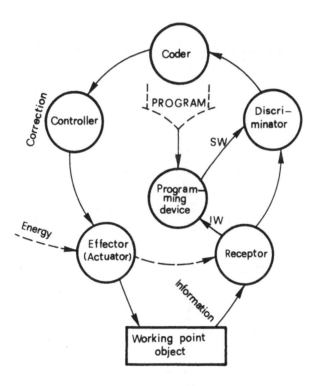

Fig. 2 Bernsteins control scheme

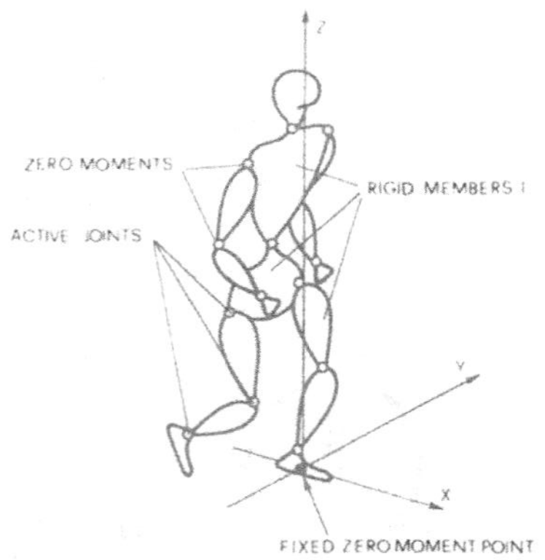

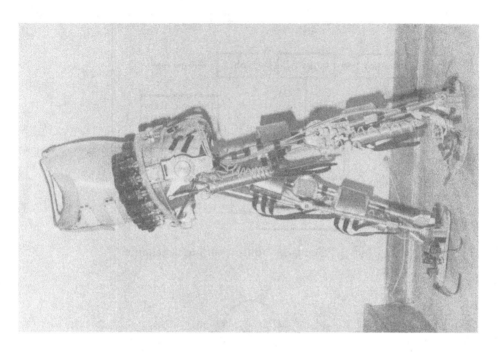

Fig. 5 Active exoskeleton for paraplegics.

Fig. 3 A human-like bipedal locomotion system

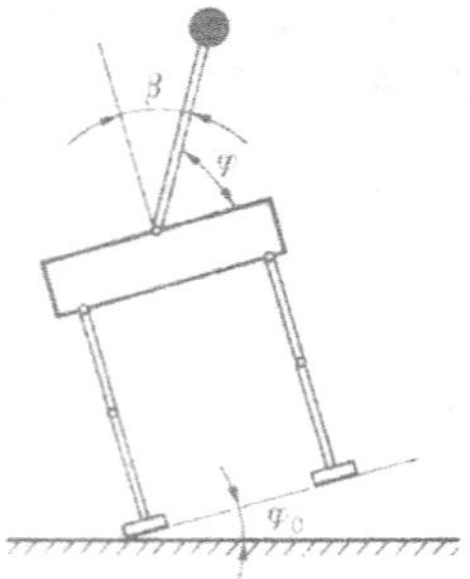

Fig. 4 Front view of the simplest locomotion system

ROBOT VISION AS A FACTOR IN THE CONTROL OF MOTION

John F. YOUNG, University of Aston,
Department of Electrical Engineering,
Birmingham, England.

(*)

Summary

Industrial robots in common use have no feedback from visual sensors. Such devices are satisfactory for use in certain applications, for example, where only coarse control is required or where mechanical disturbance is completely absent. However, the robot for general-purpose use requires a method of position-sensing feedback, and some low-cost form of visual sensing would be ideal, particularly if it could be combined with means for visual object-recognition. The difficulties and requirements of such visual sensing and recognition are reviewed, and work on various possible methods is discussed.

1. Introduction

At the present time all of the robot devices available for use in industry [1] are, in effect, blind. It is only in the case of advanced research work that there has been any attempt to produce robot devices with means for control based on vision [2, 3, 4]. The vision has in general been monocular and has incorporated standard television cameras, though sometimes complex retinas, having for example arrays of photomultiplier tubes, have been used experimentally [5].

Manipulators operated remotely by a human have been widely adopted, for example in nuclear work and in underwater work, and these have often been fitted with a television camera to signal back visual information to the operator [6].

The likely importance of vision as a factor in the future control of robot motion can perhaps be illustrated by the fact that of the 3 million nerve fibres carrying sensory information to the human brain, some 2 million are believed to carry visual information [7]. While about 1% of neurons in the human brain are devoted entirely to the visual process, as many as 60% are probably partially involved in visual perception.

Although the completely blind robot device is satisfactory in conditions where there is an extreme stability of the surroundings and where a very fine control of movement is not required, it is likely that future generations of general purpose robots will have to be fitted with some form of visual sensors.

The completely static robot device as used in industry at the present time does not need visual sensors provided that it can be firmly bolted to the machine with which it has to operate. However, if a robot device is required to be either semi-mobile or completely mobile, then visual sensing becomes desirable, if not essential. In general, the more closely the function of the robot device is required to approach or to duplicate or to replace that of the human or the animal, the more desirable does robot vision become. It is therefore worth considering briefly the development of human vision to obtain some guide to the requirements for robot vision.

2. The development of human vision.

We know something of the development of the human visual process from the experiences of people who were born blind and who had their vision restored when they were already adult; from such experiences it is known that the process of learning to see is a protracted and often painful process for the human, and it is therefore wise not to expect too much of early robot visual learning systems.

If human vision is considered from the point of view of information theory, it is found that information handling rates of about 30 bits/s are possible, but if a manual response is required then the rate falls to about 15 bits/s [8]. The eye itself is of course inherently capable of much greater rates. Human vision is affected by a number of factors such as shape, contrast, brightness, luminance of the background, sharpness of edges and viewing time. Only the central part of the retina is capable of fine discrimination or of colour vision.

While a great deal is known about the structure of the human eye at the present time, it is very unlikely that there will be any attempt to simulate the human retina directly for use in robots. This is bacause, no doubt for evolutionary reasons, the human retina is of a very poor engineering design. The sensitive photo-cells of the retina are at the back of the eye. In front of these cells, and between them and the light entering the eye, are various nerve cells and the fibres which carry away the information to the brain. The fibres have to be brought out of the back of the eye somewhere, so that they can be taken to the brain, and this is done at a single point. Since there can be no sensitive photo-sensors at this point, a blind spot is formed. In addition, because all of the light reaching the sensitive retinal cells has to pass through nerve fibres and nerve cell before it can reach the sensitive cells, it is seriously attenuated on the way.

Robot eyes will not suffer from this human deficiency. The light-sensitive cells will be facing the light and there will be no need for a blind spot.

3. Binocular vision and colour sensitivity

Most animals have two eyes, this feature having the dual function of increasing the redundancy, so reducing the risk of total visual failure in the event of damage to one eye, and at the same time of introducing the possibility of distance perception.

Except for special-purpose robots, it is unlikely that binocular vision will be very necessary in the near future, though dual optics would be very desirable from the point of view of reliability. In work on very simple mobile robot devices at Aston, we have found it useful to fit two separate single-cell visual sensors, each having a different sensitivity, since this removes the need for an iris diaphragm in preliminary work [9]. Such an approach can be thought of as akin to the use of rod and cone cells having different sensitivities in the human eye.

With binocular vision, a somewhat complex cross-coupling arrangement is required for convergence in near vision, and possibly for pupil constriction in similar circumstances. Binocular vision can be used to control head – or body-rotation for image centralisation.

The robot can, of course be fitted with more than two separate eyes, and there is no need for these to be situated in the head of the robot. For some applications, it will be and advantage if the eyes are fitted elsewhere, for example in the hand of the robot [10].

In experiments with master-slave manipulators, fitted with television viewing devices for the human operator, it has been found that stereoscopic viewing is only a little superior to two-dimensional viewing, and then only if the three-dimensional television is correctly aligned [11]. Since humans having only monocular vision can perform most tasks perfectly well, and since the feature of redundancy is not so important in the robot, where the whole eye can easily be replaced if it is damaged, it is unlikely that binocular robot vision will be regarded as important in the next generation of robots.

It is not difficult to build colour-sensitive photo-electric equipment. In the writer's experience, some of this can be quite simple, as for example colour-sorting photo-electric devices in industry. Other equipment, such as colour-television cameras, can be very complex and expensive. For many applications of robots, colour-sensitivity will be of little importance, and indeed it should be noted that as many as 10 % of humans and many animals are in fact colour-blind. However, it must be remembered that the robot can be used to extend the capabilities of the human, and robot colour vision will be most useful in this respect. For example, robot vision in the infra-red or ultra-violet regions or in other regions of electro-magnetic radiation such as X-rays will be invaluable in some cases [12]. For some applications, it will be an advantage that the robot can have an illuminating source permanently rigidly connected to the visual system, though it will be very necessary to isolate the visual detector from the heat generated by the light source. Only a thin layer of air is required for this isolation, and sensors of this form designed by the writer have been in successful use on industrial equipment for many years [13].

4. Optical illusions

The human visual system is known to be subject to various illusory effects. In studying any unknown system, it is often its peculiarities and its deficiencies which can give us a clue to its basic method of operation, and so it might possibly be worth studying the optical illusions experienced by the human.

While this seems at first to be a promising approach, it is unfortunately found that optical illusions are not an invariant experience of the whole human race. It has been found that the Zulu, in particular, does not experience many of the illusions experienced by Western man, and it has been suggested that this might be because straight lines rarely appear in his early cultural environment. On such evidence, it would appear possible that optical illusions are an acquired experience, and that they are therefore of little use in basic studies of possible optical structure in the visual system.

Recent work, for example that of Blakemore and Cooper [14], makes it appear possible that the whole development of the visual structure of the animal brain depends almost entirely upon the early environment, rather than upon hereditary factors.

The significance of such discoveries to the robotics engineer is that they indicate that an extreme of generalisation is required in any central learning system for use with robot visual systems, and at Aston [15] this generalization is being based on the known phenomenon of pavlovian association.

5. Cleanliness of robot eyes

Although the full purpose of the blinking of the eyelids of the human eye is not known, the maintenance of cleanliness is certainly one of the functions. The presence of moisture and of tears also helps in this respect.

Even though drit is known to have been a cause of failure of industrial optical pho-electric systems, it has been usual merely to rely on routine maintenance procedures, involving regular cleaning and regular replacement in order to overcome the difficulty.

It is possible to fit devices such as screen wipers and washers to photo-electric equipment for robot use, though the cost of such additions will probably make it unusual in early visual robots. It is hoped, however, that the development of simple and reliable self-cleaning photo-electric systems will be one of the beneficial effects of the widespread introduction of the robot eye. For some applications, cooling equipment or antisteaming equipment will be an additional requirement.

While eyelids are useful in the human not only for protection, for example they also make it possible to cut-off the visual scene in order to facilitate thought or sleep, it is quite possible in the robot to provide such a cut-off by purely electrical internal means. At the same time, some form of rapid-acting shutter could be most useful for protection of the robot optical system, for example against excessive ambient light. A notable case where such a rapid acting shutter system would have been most useful is the failure of the television camera on the recent human moon landing, caused by accidental exposure to direct sunlight. There is a possibility here of using the properties of various suspensions in liquids to achieve such rapid and automatic protection. There can be no doubt, however, that the human need for blinking and the psychological nature of the drives to blinking can be sometimes a big diadvantage. It has been estimated that in some cases the vision of a human can be obscured for as much as 20 % of the time by blinking [16] , and it is desirable to reduce the possibility of such extensive obscuration if robot eyes are indeed provided with this form of protection.

6. Contrast enhancement

Physiological investigations have indicated some of the features of animal methods of detection of visual images. For example, Lettvin et al [17] have found that there are various specialised nerve fibres coming from the eye of the frog, while Hubel and Wiesel [18] have discovered cells in the eye of the cat which are responsive only to movement of the image on the retina. Thus to some extent the visual detection of the nature of the image is accomplished directly at the retina, rather than at the brain of an animal.

Various technical appoaches to the detection of images are possible, but it appears that methods which emphasize the edges of the scene being viewed and which ignore areas of constant illumination are most promising for use in the robot. One possible method uses a television type of scan and incorporates delay elements so that each element of the scene being scanned can be compared with the two elements immediately above it, i.e. in the previous two scanning lines. If L_n is the output from line n of the scan, etc., then such a process can be arranged to produce an output —

$$(L_{n-1} - (L_n + L_{n-2})) k + L_{n-1} = -kL_n + (1 - k)L_{n-1} - kL_{n-2}$$

Such a system can be used vertically with a television scan, where the delay required is equal to the time-length of one line. It can also be used horizontally, where the required delay is of the order of one picture element length.

The method is similar to that described elsewhere [15].

Although television pick-up tubes seem to be ideal for use in the robot eye system, and indeed seem to have been used in most experimental visual robot systems, there are various problems. They must be carefully protected from damage such as that caused by excessive light. In general, the life of a camera tube is limited, while some tube suffer from a "burning-in" of the picture on the sensitive surface if they are exposed to a fixed bright scene for a long time.

Camera tubes are now much cheaper than was the case a few years ago, but there is still room for a cheaper system in some cases. Sometimes a robot must work in an environment where it must provide its own illumination, and in such instances the flying-spot system is worth considering. The writer produced such a system some years ago for use in a hazardous environment where the restricted life of normal tubes is a severe embarassment. A projection type of television tube produced a visual illumination raster which was projected on to the scene, while the reflected light was picked up by a photo-multiplier tube. A special phosphor was required on the scanning projection tube in order to obtain good results. The high voltages required and the poor availability of the special tubes and the standardised optical system make this scheme unsuitable for use in present-day robot devices.

An important problem with any method of vision using standard television techniques is that too much information is available, so that some information produced by the scanning system is later deliberately eliminated. It would be desirable to avoid this by the use of non-scanning methods.

7. Individual cells in arrays

While scanning methods are attractive because they make use of known techniques, the advent of large-scale integration brings the hope that low-cost, non scanning systems for robot vision will soon be available. In the projected use of these methods, one of the problems is the difficulty of taking the information away from the retina of photo-cells which would be used. In order to take out information equivalent to that from a standard television tube having a scan of only 500 lines, a minimum of 1,000 connections would be required and this is clearly quite impractical at the present time.

Consequently, it can be assumed that some of the information processing must be carried out at the retina, in order to reduce the amount of information to be taken away. It is known that the human eye is capable of an instantaneous counting of small number of objects, and it is therefore of interest to consider the possibility of introducing this facility into an artificial retina for robot use. This has been accomplished in the Aston Cybernetics Laboratory by P.S. Williams, who has used a comparison of the outputs of adjacent cells in the retina [1]. In effect, the number of edges which can be detected along any line of the visual scene is counted and then divided by two in order to determine the number of objects along that line. Such a system has limitations, and it is found to be subject to various "optical illusions", but it shows some promise, for example for the counting of cells on microscope slides.

Arising out of work on object counting, C.E. Free developed at Aston a very simple form of edge detecting retina intended for robot use [1]. Once again, the illumination of adjacent cells was compared, and an output was obtained if there was any difference in the output from the cells. It is notable that Free was able to achieve this result, and to obtain sufficient output to operate a display matrix of neon tubes, without using any transistors at all. Only resistors, capacitors and diodes were required, provided that an alternating voltage supply was available. More recent work on retinas such as these has made use of advances in integrated circuits, in particular of the "Exclusive-Or" type, to give outputs which are directly compatible with, for example, robot "brain" devices of the ASTRA type [9, 15].

It is oped that work such as this, together with developments in integrated photo-cell arrays, both unscanned and self-scanned, will have the effect of bringing down the present high cost of robot vision to the point where it can be incorporated in all forms of robot device.

8. Eye movement

In the course of most work in which attempts have been made to simulate some functions of the animal retina or eye, it has been assumed that the pattern to be viewed must be projected into a fixed position on the retina and that it must then be perceived as a whole, possibly after some processing of the output obtained.

It is known, however, that the animal eye is not at all fixed in position, and that there is always a slight tremor. In primitive creatures, this feature is used to economise on the number of retinal cells required. In more advanced animals and in man, the continous tremor appears to be just as essential to vision as in the elementary creatures. It is possible by various means to stabilise the visual image on the retina, and it is found if this is done that the picture appears to fragment into more elementary shapes such as lines, angles, curves and so on. It could be that one function of the retinal scanning motion is the coupling together of these elements in order to make integrated vision possible.

Another useful function of optical judder is the reduction or elimination of optical spacial quantisation noise. For example, it has been found possible to reduce the effective optical noise effects in light guide systems by the introduction of Judder [19], and human viewing of a picture through a visual noise overlay is facilitated if either the visual noise of the source picture is subjected to judder. Thus it seems that the introduction of judder into robot vision systems is worth considering, and this could possibly be applied electrically, directly at the retina.

The other method of introduction of optical judder is by the use of light-conducting fibres carrying the light to the photo-detectors. The end of the fibres can then be vibrated. This method has additional advantages, for example it makes it possible to use a very small effective retina with a much larger physical array of photo-detectors, so greatly simplifying the problem of electrical connection to the photo-detector array. Another potential advantage is that other fibres in the same bundle can be used to carry illuminating light which is then reflected from the object being observed into the sensing fibres. It is believed that the use of glass rods as light fibres was suggested as long ago as 1926 by Baird, though it is only since the advent of fine fibres, in some cases coated with a layer of low refractive index, that the method has become important.

9. The eye following a moving object

When the human eye looks at an object which is not moving in the field of vision, the eye moves in a series of rapid jerks, each followed by a period of rest during which the scene is viewed. This method is used, for example, by the human when reading printed matter. In some cases when a moving object is being viewed, the whole head moves in order to keep the image fixed on the retina. However, if the head is restrained from moving, then the mode of action becomes more complex.

If the movement of the object is predictable, as for example with a pendulum, then the control system of the eye appears to apply a phase lead of the order of 10%. If the object being viewed moves more rapidly, then the phase shift rapidly increases [20]. Such investigations can help to suggest methods for use in robot vision control.

If there is a remote human operator, the viewing camera can be arranged to follow the movements of the head of the operator directly [21]. Such methods show promise both for investigation of the required control characteristics and for use, in conjunction with electronic learning systems, in the course of teaching the robot control to carry out various tasks.

10. Automatic focussing of the robot eye

Various methods of automatic focussing which would be suitable for use in the robot eye have been tried. The simplest approach is to use a pinhole form of small-aperture lens, though this is not suitable for general-purpose use. In one equipment, a lens was moved by a motor through a gearbox, the motor being controlled by a peak-holding circuit operated from the signal obtained from a photo-multiplier cell [22].

A promising form of lens for robot use is that having one flexible wall, the curvature of which is altered by pumping in a liquid [23]. Another interesting approach is the use of an infra-red beam, the reflection from which is focussed [24]. This automatically focusses the visible light reflected from the object at the same time. The light spot is about 15 cm in diameter at a distance of about 20 m from the camera and it is chopped at a rate between 20 Hz and 70 Hz. The advantage of this approach is that a very small portion of the scene to be viewed is automatically separated out for focussing.

An approach to focussing used at Aston by R.J. Davies used a perturbation technique, the perturbation being obtained by the rotation of a plastic disc in front of the camera tube which was itself moved for focussing by a fine thread screw driven by a motor [1].

11. Persistence of vision

The persistence of human vision leads to optical illusions which are almost taken for granted. If it was not for this phenomenon, films and television would not be possible.

Now in the case of the robot, persistence of vision is optional, and indeed it would be difficult to obtain an exact match to the human characteristics under all conditions of lighting etc. In some cases, for example in the observation of high-speed phenomena, it will be an advantage that a robot can have a more rapid visual response than a human, and in most applications it will be advantageous if the robot is fitted with a standardised rapid response visual system, any required slowness of reaction being obtained by operation on the electrical output.

12. Conclusions

It can be seen from the above brief review that while a great deal is known about methods of achieving successful robot vision, and about the requirements for various robot applications, it is not practicable at the present stage to design a standardised visual system suitable for all purposes. At the same time, unless standardised methods are produced, it will not be possible to take advantage of the economy of scale and so to introduce robot vision on a wide scale.

It would seem, then, that work on the production of standardised and economical visual systems for robot use is an urgent necessity if the next generation of robots, incorporating vision, is not to be so expensive that it cannot be widely used.

The advantages to be obtained by the addition of simple and reliable visual systems to the robot are such that whoever wins the race to develop a low-cost approach will have the chance of capturing a large share of the potential market for general-purpose domestic and industrial robots.

REFERENCES

[1] YOUNG, JOHN F., "Robotics," Butterworth (U.K.) and John Wiley (U.S.), 1973.

[2] SUTRO, L.L., and McCULLOCH, W.S., "Steps toward the automatic recognition of unknown objects," Proc. I.E.E./N.P.L. Conf. Pattern Recog., 1968.

[3] FELDMAN, J.G., et al. "The Stanford hand-eye project," Proc. Int. Conf. Art. Intell., N.Y. (A.C.M.), 1969, p.509.

[4] MICHIE, D., et al., "Vision and assembly as a programming problem," Proc. Conf. On Industrial Robot Technol., Nottingham, 1973, March, p. 185.

[5] TAYLOR, W.K. "Machines that learn," Sci. J., 4, 1968, Oct., p. 102.

[6] GOERTZ, R., et al. "Preliminary report on the ANLmark4A master-slave manipulator," Proc. 14th Conf. Rem. Systems. Technol. (A.N.S.), 1966.

[7] YOUNG, JOHN F. "Information Theory," Butterworth, 1971.

[8] LICKLIDER, J.C.R., et al. "Studies in speech, hearing and communication," Report M.I.T. Acoutics Lab., 1954.

[9] YOUNG, JOHN F. "Cybernetics," Iliffe (U.K.) and Elsevier (U.S.), 1969.

[10] HEGINBOTHAM, W.B., et al., "The Nottingham SIRCH assembly robot," Proc. Conf. On Industrial Robot Technol., Nottingham, 1973, p.129.

[11] GOERTZ, R., Manipulator systems development at A.N.L., Proc. 12th Conf. "On Remote Systems Technol.," (A.N.S.), 1964, Nov., p. 117.

[12] MORTEN, F.D.,"Infra-red detectors and their applications," Mullard Tech. Comm., 10, 1968, May, p.75.

[13] MAWER, J.C.,"Electronic colour registration in printing," Industrial Electronics, 1, 1963, Aug., p. 561.

[14] BLAKEMORE, C., COOPER, G.F.,"Development of the brain depends on visual environment," Nature, 228, 1970, p. 447.

[15] YOUNG, JOHN F.,"Cybernetic Engineering," Butterworth (U.K.) and John Wiley (U.S.), 1973.

[16] LAWSON, R.W.,"Blinking, its role in physical measurements," Nature, 161, 1948, p. 154.

[17] LETTWIN, J.Y. et al.,"What the frog's eye tells the frog's brain," Proc. I.R.E., 47, 1959, p. 1940.

[18] HUBEL, D.H., WIESEL, T.N.,"Receptive fields, binocular interaction and functional architecture in the cat's visual cortex," J. Physiol., 160, 1962, p. 106.

[19] BALLANTINE, J.M., ALLEN, W.B.,"Fibre optics," Sci. J., 1, 1965, Spt., p. 1.

[20] SUGIE, N., JONES, G.M.,"A model of eye movements induced by head rotation," Trans. I.E.E.E., SMC1, 1971, July, p. 251.

[21] GOERTZ, R.C., et al.,"The ANLmkTV2-an experimental 5-motion, head-controlled television system," Proc. 14th Conf. Remote Systems Technol., (A.N.S.), 1966, p. 124.

[22] DEELEY, E.M., ALLOS, J.E.,"Automatic focussing of an optical system by extremum control," Proc. I.E.E., 114, 1967, Jan., p. 161.

[23] ANON.,"Variable-focus lenses," Control, 12, 1968, May, p. 469.

[24]   ODONE, G., Camera's infra-red eye focuses on new vistas for ranging, Electronics, 43, 1970, April 27th, p. 10.

SURVEY PAPERS

INDUSTRIAL ROBOTS
PROBLEMS OF DESIGN AND APPLICATION

Hans-Jürgen WARNECKE, Full Professor,
University of Stuttgart,
Institute of Production and Automation
Stuttgart, Federal Republic of Germany

(*)

Summary

Industrial robots are programmable machines with several degrees of freedom, with grippers or special handling attachments, designed for application in industry. The development started in 1954 and there are many social, technical and economical reasons for a growing interest in these units, it is expected that they will give a push to further automation in batch production. There are already many types on the market. The characteristics of the known industrial robots are collected and given in this paper. On the other hand the results of a first investigation of working places are told which give a first impression which demands an industrial robot has to fulfill. Two tendencies can be observed one is the very sophisticated type with sensors for pattern recognition and the other one is a modular system for optimal adaption without sophisticated sensors when the workpieces are fed in an accurate position.

(*) All figures quoted in the text are at the end of the lecture.

1. Survey

The word "robot" is not yet clearly defined. Our imagination is impressed by science-fiction stories, where the robot is androide system capable of learning and moving. A questioning of experts, which of different sets given for example they would call robot, has had the result shown in fig. 1. [1]. A broad variety of meanings can be seen, although as expected the majority sees automated learning and moving systems as robots. The practical application of such units in near future is mainly seen in industry, in data and information processing, in administration and in surroundings hostile to the human being for instance in space. Fig. 2 shows the result of a questionaire by the Delphi-method. Capabilities of organizing are generally not expected before the year 2000, whereas robots with capabilities of manipulating and sensoring will be applicated in this decennium and developed intensively. These units shall be named industrial robots in the following and they are defined in that way:

Industrial robots are programmable machines with several degrees of freedom, with grippers or special handling attachements, designed for application in industry.

The development of industrial robots started in USA in 1954, the first units came on the market in 1962. In the last 5 years a more intensive development started in USA, Europe and especially in Japan.

The reasons for the great interest are manifold. The main reasons are [2,3].

Economic reasons:

- growing competition on the world market
- rising labour costs
- Capital intensive manufacturing equipment compels maximum efficiency round the clock.

Especially may be said the labour shortage for jobs in hostile surroundings. The experiences of the manufacturers of industrial robots as yet show that at these places the user sees the necessity in the first step and is ready for application. In these cases legislation will be pacemaker more and more. In USA for instance 1974 a law will forbid working at machines which operate with shutting tools. Also the allowed noise level will be lowered and compel new solutions.

Another pacemaker is the said necessity for the use of expensive capital goods, especially numerical controlled machine tools and manufacturing systems,. An economic application is only given when there is an output round the clock. For shift work no labour is available and we must find automatic solutions.

Social reasons:

- Keeping away man from monotonous work
- making man free for creative work
- labour shortage
  for work of low value
  for shift work
  for work under bad conditions as high temperatures, noise and dirt
- less working hours
- more safety at work
- higher standard of living
- lost bond of man to his working place and the manufactured product.

Technical reasons:

- The human efficiency is limited in working accuracy
- velocity
- load capacity
- perseverance and steadiness

2. System Production

In the following problems of design and application of industrial robots in the system "production" shall be shown. The fundamental structure is explained in fig. 3, The subsystems are:

- production planning for once fixing all technical things as operations, machines and tools.
- production control for quantities and dates.
- transportation system for material and tools.
- material handling system for feeding, orientating and positioning of workpieces and tools in the range of the machine tools.
- working system for forming the material by energy and information.

- quality system for securing the quality of the products.
- control system for all the systems.
- energy supply system for energy supply.

The demands to the human being in the system "production" have steadily changed unitl now by technical developments. The compelte freedom of manual operations and of control functions in the range of the working system and relieve of men also in the planning, controlling and inspecting ranges of the system production has been achieved in mass production. We aim at a factory which makes possible a maximum in productivity and quality with a minimum of direct bond of the human being to the production process.

In batch production we are far from that situation. Emphasis of rationalization lies therefore today on batch production of workpieces. By the tendencies to adapt the offer to the individual whishes of the customer — we have a customer market — batch production will gain more importance. Their rationalization is important for further economic development.

3. Handling System

Handling of tools and especially of workpieces in the system production has the lowest level of automation in batch production. This is due to the volume and variety of problems in manufacturing industry because every designed work piece has — generally speaking — as many shapes as operations are necessary during its manufacturing process. At each stage the workpieces get another shape or other characteristics which must be considered in handling. The more the work pieces reach the final stage the more difficult generally is their handling, highest in assembly operations. Whereas in part production only equal workpieces out of the same material have to be fed in determined time intervals, in assembly several workpieces of different shape, dimension and stiffness must be handled at one time.

In handling of tools the variety is not so big, but fundamentally the same considerations are valid.

Supposition for an improvement in handling of workpieces is a systematic analysis of all operations. In spite of the said variety certain operations are repeated in all these tasks. Always in automation of the handling tasks workpiece, machine tool and possible handling equipment must be considered at the same time. Only if all three components are designed for automation and looked on by the methods value engineering an optimum solution will be achieved (Fig. 4). In analysing it is sensible to use symbols for instance laid down in VDI-Richtlinie 3239.

A main disadvantage of all industrial robots on the market is missing of all sensoring functions. It is not yet clear whether in furture it is more economic to create at the working place the possibilities for application of blind and unfeeling industrial robots for instance by orientations and positioning of the workpieces for handing or to develop optical and feeling sensors. The speed and accuracy of the human sensors and his capability of combination and decision-making will be not achieved by technical solutions in next future. Therefore the application of industrial robots is depending on the exact feeding of the workpieces. The working operations must run automatically and must not influence the further handling process that means the workpiece must be in an exact position after the working operation. The machine tool must be linked to the industrial robot in its control and in many cases some more functions of the machine must be automated for instance clamping of the work piece.

4. Design of industrial robots

The fundamental structure and the motion axis are shown in fig. 5 and 6. There is a combination of known technics for drive, cinematics, control, measurement, gripper and sensors. A great number of models are already offered (Fig. 7), mainly from Japan, which develops in a very intensive way [4]. At present about 2500 units are in application worldwide, although this is a very rough guess and the not yet approved definition has an influence.

About 70 per cent of the offered models work in cylindrical coordinates, that means with two translation- and one rotational motion. In addition there is a motion of the gripper, so that four degrees of freedom present (Fig. 8). More degrees of freedom are necessary for complicated movements for instance when spot welding, painting or assembling. In this analysis the motion of the fingers of the gripper and a motion of the whole unit are not counted.

Driving-, control- and measuring systems are shown in fig. 9 and 10 [5]. The hydraulic drive is eminent, because any position in the range can be set and high forces can be gained with small pistons. Hydraulic units do not only allow point-to-point control (PTP) but also continuous path control (CP). About 90 per cent of the industrial robots have PTP-control.

The simple and cheap pneumatic drive has as yet the disadvantage, that he only allows setting of some positions. In the meantime there are solutions known working with brakes and length pickups in order to get any wanted accurate stop. This new technical solution was caused by industrial robot application.

In future more and more electrical drives will come to the market because of the development of new motors allowing economic designs by combining drive and control.

Nearly 60 per cent of the offered models have no length — and angle — measurement, but work against fixed stops. If there is a measuring system they are analogue or digital. The positioning accuracy is below ± 1 mm for more than 50 per cent of the units (Fig. 11). That is sufficient for most cases, not for many assembly operations. The highest accuracy give fixed stops but then flexibility is poor. Accuracy is influenced very much by speed of motion and moving masses. Speeds are on the average between 500 and 1000 mm/s or 90 to 180 °/s. Therefore many models are too slow in comparison to human speeds when moving small parts. Here further development is necessary. Load capacities are shown in Fig. 12. Many models carry up to 10 kg. a weight very often in machine industry. There are models on the market with much higher load capacities but less speeds and accuracy.

Programming of industrial robots is performed directly on the spot of application by doing all operations manually and setting the program. That case is only given for about 17 per cent of the types. About half of the models work with fixed or settable stops and setting the operating process in a plug board. CP-control with magnetic tape or magnetic drum allows direct programming by once performing the handling operation manually. Afterwards it can be repeated automatically always.

5. Demands of the working place

Inspite of the many types industrial robots are coming into the market very slowly. Reasons are insufficient information of the potential user, unclarified conditions at the working places and little experience refering especially to maintenance, reliability and life time. It is necessary to investigate the necessities of the working places systematically in order to inform manufacturer and user accurately.

Fundamental conditions for the application of todays industrial robots are simple and often repeated motions with one arm. The accuracy of positioning don't have to be too high and sensoring functions not too difficult. Inspection tasks the industrial robot cannot perform generally besides programmable and sensorable decisions. The big advantage is that the industrial robot is insensible against impact of the surroundings and its flexibility incomparison to specialized automated feeding equipment, that means its adoption to new handling tasks.

Fig. 13 shows the result of an investigation of 173 working places in the parts production of an automobile factory [6]. From that the following characteristics must be met:

a) by more than 80 per cent
    one arm
    PTP-control
    X-motion of the arm
    Z-motion of the arm
    C-motion of the arm with 180 °

b) about 50 ... 80 per cent
    positioning accuracy = 1 mm
    X = 1000 mm
    5 positions in X- and C-direction
    A-motion or
    B-motion or
    C-motion of the gripper

Besides that the industrial robot must be set fast and simple, a condition for economical automation of many working places. Another point is that the industrial robot must not take much area for instance by separating the mechanical system from the control system.

6. Future development

The future development of industrial robots will go two ways, both will have their market. The one development will go to more sophisticated types with sensors permitting optical pattern recognition and simple replacing the worker at a working place or machine. On the other hand one will try to change the total manufacturing system to a simpler adoption to the industrial robot in avoiding sophisticated functions and having the workpieces in an order at any time. That will lead to a modular system the elements of which meet the several demands of working places. Both directions of development can already be recognized.

Similar considerations are apt to grippers. It is in many cases more sensible to have a special gripper because a universal one copying the human hand is too complicated and expensive.

In the next years a strongly growing market will arise. Many tendencies in the social fields and the raises in wages will push the application. In many cases the economic side where a pay-back time of 1 to 2 years is not yet reached, will be of interest in near future. Which functions shall be automated by which means must be decided in every individual case. Automation always in investment and a longtime bond of capital and "over-automation" is a risk. An important condition for automation is a high reliability of all components. The more sophisticated and expensive the equipment becomes the higher are the demands for reliability and for taking precautions to minimize costs of breaking-down. Sinked machine-tools for instance must have magazines between it possible. Another demand is that for easy operating and maintenance.

REFERENCES

[1]   N.N., Delphi-Befragung der Industrie-Anlagen — Betriebsgesellschaft mbH., Ottobrunn.

[2]   WARNECKE, H.J., SCHRAFT, R.D., Industrie-Roboter Krausskopf-Verlag, Mainz, 1973.

[3]   WARNECKE, H.J., Betrachtungen zur Automatisierung der Handhabung. Referate des 3. Internationalen Symposium über Industrieroboter 29. bis 31. Mai 1973, Zürich, Verlag moderne Industrie, München, 1973, pp. 13 - 26.

[4]   HASEGAWA, Y., Analysis and Classification of Industrial Robot characteristic, see 3, pp. 53 - 70.

[5]   SCHRAFT, R.D., Der Aufbau von Industrierobotern — eine Untersuchung des weltweiten Angebotes, see 3, pp. 71 - 82.

[6]   HERRMANN, G., Anforderungen an Industrieroboter in der Fertigung, see 3, pp. 167 - 176.

No.	Example	Score
1	Mobile system with the capability to "learn" and to find the shortest path across an area with arbitrarily positioned obstacles to a predetermined goal without collision.	120
2	Mobile system with the ability to collect cubes which have been randomly distributed within an area	87
3	ELSIE (electro light sensitive with internal and external stability) also known as "machina speculatrix" or simply "electronic tortoise"	73
4	"Lunochod" the Russian vehicle for lunar surface exploration	49
5	System for machine translation taking grammar and syntax full into account	40
6	Electromechanic artificial limb controlled by the human nervous system	10
7	System for the continuos measurement of vehicles distance (cars, trains, etc.) ensuring a minimal distance	9
8	Checking system for money or stamps against counterfeits	7

Fig. 1   What is a robot ?

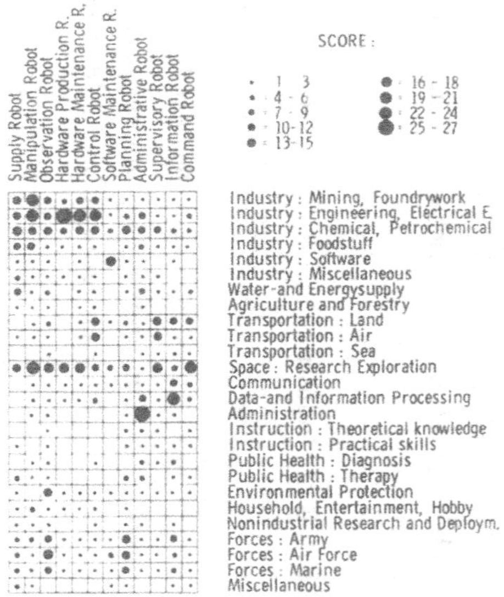

Fig. 2   Result of a Delphi-inquiry to robot application

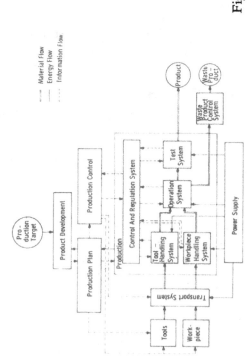

Fig. 3   System Production

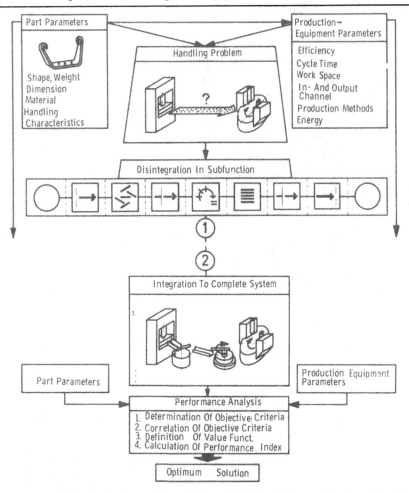

Fig. 4   Systematic solution of a handling problem

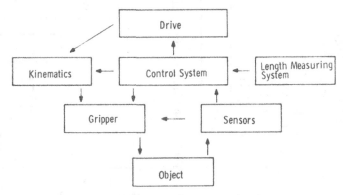

Fig. 5   Structure of industrial robots

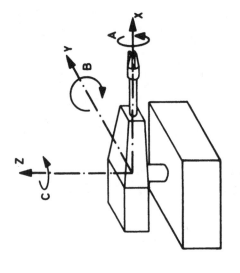

Fig. 6   Axis of motion of an industrial robot

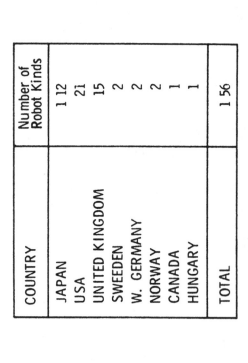

COUNTRY	Number of Robot Kinds
JAPAN	1 12
USA	2i
UNITED KINGDOM	15
SWEEDEN	2
W. GERMANY	2
NORWAY	2
CANADA	1
HUNGARY	1
TOTAL	1 56

Fig. 7   Number of types of industrial robots by 4

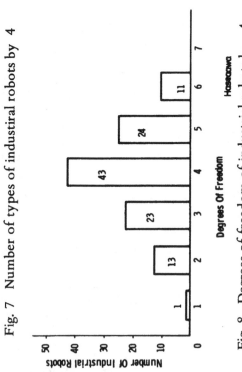

Fig. 8   Degrees of freedom of industrial robots by 4

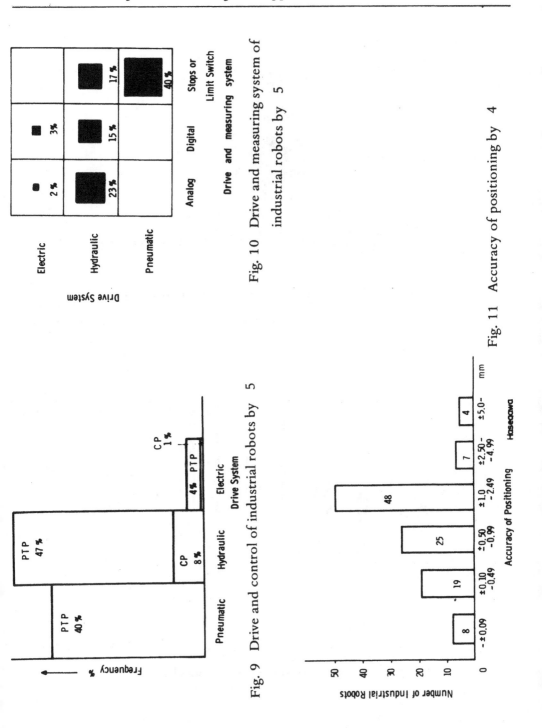

Fig. 10 Drive and measuring system of industrial robots by 5

Fig. 11 Accuracy of positioning by 4

Fig. 9 Drive and control of industrial robots by 5

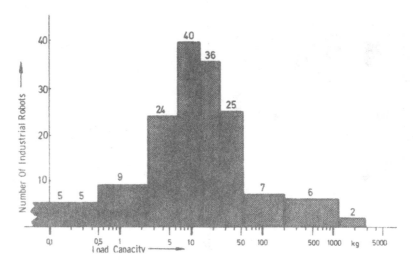

Fig. 12   Load capacity of industrial robots by   5

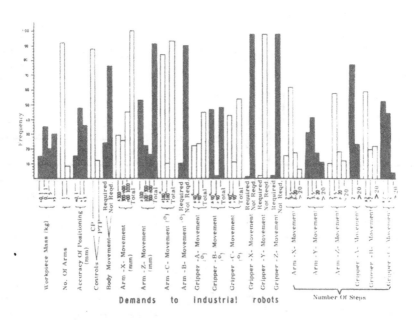

Fig. 13   Demands to industrial robots by   6

CONTENTS

Printed in the United States
By Bookmasters